Novel Feed Ingredients: Improving Health Status, Milk and Meat Quality in Small Ruminants

Novel Feed Ingredients: Improving Health Status, Milk and Meat Quality in Small Ruminants

Editors

Panagiotis Simitzis
Athanasios I. Gelasakis

MDPI • Basel • Beijing • Wuhan • Barcelona • Belgrade • Manchester • Tokyo • Cluj • Tianjin

Editors

Panagiotis Simitzis
Agricultural University of Athens
Greece

Athanasios I. Gelasakis
Agricultural University of Athens (AUA)
Greece

Editorial Office
MDPI
St. Alban-Anlage 66
4052 Basel, Switzerland

This is a reprint of articles from the Special Issue published online in the open access journal *Animals* (ISSN 2076-2615) (available at: https://www.mdpi.com/journal/animals/special_issues/feed_health_milk_meat_small_ruminant).

For citation purposes, cite each article independently as indicated on the article page online and as indicated below:

LastName, A.A.; LastName, B.B.; LastName, C.C. Article Title. *Journal Name* **Year**, *Volume Number*, Page Range.

ISBN 978-3-0365-7154-6 (Hbk)
ISBN 978-3-0365-7155-3 (PDF)

Contents

About the Editors

Panagiotis Simitzis

Panagiotis Simitzis works as an Associate Professor of Animal Husbandry, Ethology and Welfare at the Department of Animal Science of Agricultural University of Athens. He holds a PhD in animal husbandry, behavior and products' quality from the School of Animal Biosciences, Agricultural University of Athens, and a BSc from the same school. His research interests include Applied Animal Behaviour Science, Animal Welfare, Animal Husbandry, Quality Characteristics of Animal Products and Functional Foods. He has published more than 65 scientific articles in international journals, 10 book chapters and participated in national and international conferences with more than 70 oral or poster presentations.

Athanasios I. Gelasakis

Athanasios I. Gelasakis is an Assistant Professor of Farm Animal Diseases at the Department of Animal Science of Agricultural University of Athens. He holds a PhD in animal management, health and welfare from the School of Veterinary Medicine, Aristotle University of Thessaloniki, and a BSc from the same school. He received a MSc in Public Health from the National School of Public Health in Athens and he is a diplomate of the European College of Small Ruminant Health Management (Dipl. ECSRHM). He has participated in 31 national and international research projects and has published as an author 65 publications in peer-reviewed journals, three book chapters and 145 publications in conference proceedings. He has extended experience in farm animal health and welfare, infectious and non-infectious diseases, husbandry, herd health management and nutrition and their association with productivity, and sustainability of farms.

Preface to "Novel Feed Ingredients: Improving Health Status, Milk and Meat Quality in Small Ruminants"

Alternative feeding strategies are continuously employed in small ruminant production systems in an effort to fortify animal health and welfare status and improve milk and meat quality characteristics, leading to the prolongation of their shelf life and to an increase of their marketable value. Incorporation of novel feed ingredients into small ruminants' diets has therefore been highlighted as an effective approach for the enhancement of milk and meat intrinsic quality since bioactive compounds are preferably deposited where they are mostly required. Dietary supplementation with these novel additives could manipulate bacteria involved in ruminal biohydrogenation, decrease methane emissions, enhance animals' health and well-being status, reinforce antioxidant and anti-spoilage properties and positively modify milk and meat quality characteristics. The present Special Issue includes original research and reviews on the effects of novel feed ingredients and their bioactive compounds on health and welfare status, as well as milk and meat intrinsic quality of small ruminants. We hope that this Special Issue will contribute to enlarge our current knowledge about the novel feed ingredients used in small ruminants and establish their regular use.

Panagiotis Simitzis and Athanasios I. Gelasakis
Editors

animals

MDPI

Article

The Impact of Whole Sesame Seeds on the Expression of Key-Genes Involved in the Innate Immunity of Dairy Goats

Christina Mitsiopoulou [1], Kyriaki Sotirakoglou [2], Dimitrios Skliros [3], Emmanouil Flemetakis [3] and Eleni Tsiplakou [1,*]

1 Laboratory of Nutritional Physiology and Feeding, Department of Animal Science, School of Animal Biosciences, Agricultural University of Athens, Iera Odos 75, 11855 Athens, Greece; chr_mitsiopoulou28@hotmail.com
2 Laboratory of Mathematics and Statistics, Department of Natural Resources Management and Agricultural Engineering, Agricultural University of Athens, Iera Odos 75, 11855 Athens, Greece; sotirakoglou@aua.gr
3 Laboratory of Molecular Biology, Department of Biotechnology, School of Food, Biotechnology and Development, Agricultural University of Athens, Iera Odos 75, 11855 Athens, Greece; dsklhros@gmail.com (D.S.); mflem@aua.gr (E.F.)
* Correspondence: eltsiplakou@aua.gr; Tel.: +30-210-529-4435; Fax: +30-210-529-4413

Simple Summary: This study examined the impact of whole sesame seeds (WSS), rich in both linoleic acid and lignans, on the innate immunity of goats. WSS were incorporated in the concentrates of the control group at 5 and 10% respectively, by partial substitution of both soybean meal and corn grain. The highest supplementation level of WSS resulted in a significant down-regulation in the expression levels of several pro-inflammatory genes in the neutrophils of goats. In conclusion, the dietary supplementation of goats with WSS might be a good nutritional strategy to improve their innate immunity.

Abstract: Whole sesame seeds (WSS) are rich in both linoleic acid (LA) and lignans. However, their impact on the innate immunity of goats is not well studied. Twenty-four goats were divided into three homogeneous sub-groups; comprise one control (CON) and two treated (WWS5 and WWS10). In the treated groups, WSS were incorporated in the concentrates of the CON at 5 (WSS5) and 10% (WSS10) respectively, by partial substitution of both soybean meal and corn grain. The expression levels of *MAPK1*, *IL6*, *TRIF*, *IFNG*, *TRAF3*, and *JUND* genes in the neutrophils of WSS10 fed goats were reduced significantly compared with the CON. The same was found for the expression levels of *IFNG* and *TRAF3* genes in the neutrophils of WSS5 fed goats. Both treated groups primarily affected the MYD88-independent pathway. The dietary supplementation of goats with WSS might be a good nutritional strategy to improve their innate immunity.

Keywords: whole sesame seeds; innate immunity; blood; neutrophils; goats

Citation: Mitsiopoulou, C.; Sotirakoglou, K.; Skliros, D.; Flemetakis, E.; Tsiplakou, E. The Impact of Whole Sesame Seeds on the Expression of Key-Genes Involved in the Innate Immunity of Dairy Goats. *Animals* **2021**, *11*, 468. https://doi.org/10.3390/ani11020468

Academic Editor: Roberto Rasero
Received: 12 January 2021
Accepted: 8 February 2021
Published: 10 February 2021

1. Introduction

N-3 polyunsaturated fatty acids (PUFA) in humans have anti-inflammatory role [1] since resolve inflammation [2] and eliminate pain in inflammatory circumstances [3]. On the other hand, increased consumption of linoleic acid (LA), the main fatty acid (FA) of the n-6 PUFA group, might be related with inflammatory diseases due to its metabolization in LA-derived pro-inflammatory lipoxins and arachidonic acid, which further leads to pro-inflammatory eicosanoids and prostaglandins production [4]. An enhancement in the concentrations of pro-inflammatory leukotriene and prostaglandins [5] was found in rats fed with high LA diets. Significantly higher tumor necrosis factor a (TNFA), and interleukin-7 concentrations in the liver of pregnant rat, consumed a high compared with low LA diet, was observed, without the cytokines content in their blood to be affected [6]. Similarly, excessive dietary LA consumption increased significantly the TNFA content in plasma and nuclear factor-kappa B (*NF-KB*) expression in rats' aortas [7]. On the other hand, recent reviews and meta-analysis studies provide evidence that LA intake decreases [8] or has no effect on cardiovascular diseases [9,10] disputing its role in chronic diseases involving inflammatory process.

So far, to the best of our knowledge, no information exists on the impact of LA in the innate immunity of productive animals, and particularly in goats. Thus, whole sesame seeds (WSS), due to their high LA (44%) content [11], can be used in goats' diets to test this hypothesis. Moreover, WSS contain lignans such as sesamin and sesamolin, which might have several beneficial effects in immunity [12]. The anti-inflammatory properties of sesame in rats' models through in vitro and in vivo trials have been reviewed recently [13]. Sesamin down-regulates the expression of Toll-like receptor 4 (*TLR4*) gene in lipopolysaccharide (LPS) stimulated BV-2 microglial cell line of rats, in a dose dependent matter in vitro [14]. Accordingly, 50 μM of sesamin suppressed the activation of p38 mitogen-activated protein kinase (MAPK) signaling pathway after its stimulation with LPS [15]. A significant decline in the expression of interleukin 1 Beta (*IL1B*), interleukin-2 (*IL2*) and *TNFA* genes in mouse senescence-accelerated brain cells was found, when fed with sesaminol [16]. So far, the impact of dietary inclusion of WSS in the immune system of produced animals has not been studied.

The immune system is broken down into innate and adaptive [17]. Neutrophils comprise one of the main cellular component of the innate immune system and the first line of defense against pathogens [18]. The innate immune system employs special receptors known as pattern-recognition receptors (PRRs) such as NOD-like receptors (NLRs) that recognize pathogen- or damage-associated molecular patterns (PAMPs and DAMPs, respectively) [19]. Among the PRRs, Toll-like receptors (TLRs) enact the induction of immune response [20]. All TLR signaling pathways end up in activation of the transcription factor nuclear factor-kappa B (*NF-KB*) and interferons (IRFs), which regulate the outcome of innate immune responses [19]. In addition, a core element of the NF-KB cascade is the IκB kinase (*IKK*) complex or conserved helix-loop-helix ubiquitous kinase (CHUK) which is encoded by the *CHUK* gene [21]. TLR activation stimulates the release of various inflammatory cytokines (*TNFA*, interferon *IFNG*, interleukins (*IL1B*, *IL2* and *IL6*) and immune modulators such as IL8, C-C motif chemokine ligand 5 (*CCL5)* and chemokine (C-X-C motif) ligand 16 (CXCL16) [22–24]. After that, *IL6* induces downstream signaling of the signal transducer and activator of transcription 3 (*STAT3*) [25]. Upon PAMPs and DAMPs recognition, TLRs recruit Toll-interleukin-1 receptor TIR domains, which transmit downstream signals via adaptor molecules such as myeloid differentiation primary response gene 88 (*MYD88*) and the TIR (Toll/Interleukin-1 Receptor) domain-containing adaptor protein inducing interferon beta (TRIF) [26,27]. The MYD88-dependent pathway activates the *IRF5* gene expression [28] and the pathway involving mitogen-activated protein kinases (MAPKs). TNF Receptor-associated Factor 3 (*TRAF3*) is incorporated into both MYD88 and TRIF complex, activating MYD88-dependent signaling and suppressing TRIF-dependent pathway [29]. TRAF3 mediates activation of *IRF3* [30]. The extracellular signal regulated kinase (ERK)–mitogen-activated protein kinase pathway determines the regulation of *JUND* gene expression [31]. Finally, Heme Oxygenase-1 (*HO1*) gene has the ability to modulate immune responses [32].

Taking into account all the above, the objective of this study was to investigate the effects of dietary inclusion of WSS at two different levels (5 and 10%) on the expression of selected key-genes (*NLRC3, TLR4, MYD88, NF-KB, MAPK1, IL1A, IL1B, TNFA, TNFB, IL2, IL6, IL10, STAT3, TRIF, IRF3, IFNG, TRAF3, IRF5, CCL5, IL8, CXCL16, HO1, JUND* and *CHUK*) involved in the innate immunity of dairy goats.

2. Materials and Methods

2.1. Animals and Diets

Animal handling procedures were performed in accordance with protocols approved by the Agricultural University of Athens Ethical Committee of the Faculty of Animal Sciences. Twenty-four goats were divided into three homogenous subgroups ($n = 8$) according to their fat-corrected milk yield (1.00 ± 0.22 kg/day) and body weight (44.9 ± 5.4 kg). The goats were fed on a group basis with a basal diet consisted of alfalfa hay, wheat straw and concentrates (Forage/Concentrate ratio = 50/50), for a seven-day adaptation period. The forages were provided separately from the concentrates while they were both offered to the animals twice a day (in two equal parts at 08:00 and 18:30 h) after milking. After the adaptation period the control goats continued to consume the basal diet, in the concentrates of which hulled sesame seeds were not included (CON). On the other hand, in the concentrates of the two other groups whole sesame seeds at 5 (WSS5) and 10% (WSS10) respectively, were incorporated by partial substitution of both soybean meal and corn grain (Table 1), in order the dietary treatments to be iso-energetic and iso-protein, and to meet the animals' average maintenance and lactation requirements [33]. The quantities of food offered to the animals were adjusted every two weeks, according to their average requirements, based on their body weight and milk fat-corrected yield. Diet selectivity did not occur, and no refusals of forage and/or concentrates were observed. The mineral and vitamin premix of both concentrates contained the following (per kg as mixed): 150 g Ca, 100 g P, 100 g Na, 100 mg Co, 300 mg I, 5000 mg Fe, 10,000 mg Mn, 20,000 mg Zn, 100,000 mg Se, 5,000,000 IU retinol, 500,000 IU cholecalciferol and 15,000 mg α-tocopherol. The experimental period, lasted 100 days and all the animals had free access to fresh water.

Table 1. Nutrients and fatty acids intake from forages and concentrates, and the total antioxidant capacity and phenolic content of concentrates only.

Daily Nutrients Intake (g/goat)	**Diets (Forages and Concentrates)**		
	CON [1]	**WSS5 [2]**	**WSS10 [3]**
Dry matter	2028.4	2027.6	2035.6
Ash	143.4	147.6	153.3
Ether extract	44.7	72.7	100.9
Crude protein	323.6	323.2	334.6
NDF [4]	766.1	795.1	782.7
ADF [5]	504.1	518.9	513.5
Daily Fatty Acids Intake (g/goat)			
$C_{14:0}$	0.34	0.34	0.33
$C_{15:0}$	0.13	0.14	0.14
$C_{16:0}$	7.96	10.27	14.06
$C_{16:1(n-7)}$	0.25	0.29	0.34
$C_{17:0}$	0.26	0.27	0.21
$C_{18:0}$	1.70	2.52	4.48
$C_{18:1(n-9)}$	8.70	20.18	31.19
$C_{18:2(n-6)c}$	20.28	33.42	44.58
$C_{20:0}$	0.15	0.22	0.34
$C_{18:3(n-3)}$	3.77	3.87	3.94
$C_{20:2}$	0.10	0.09	0.10
$C_{22:0}$	0.31	0.34	0.39
$C_{23:0}$	0.10	0.10	0.10
$C_{22:2}$	0.01	0.01	0.01
$C_{20:5(n-3)}$	0.04	0.04	0.04
$C_{24:0}$	0.46	0.48	0.50
$C_{24:1(n-9)}$	0.14	0.14	0.14
		Concentrates	
Total Antioxidant Capacity	CON [1]	WSS5 [2]	WSS10 [3]
FRAP [6] (μM ascorbic acid/g DM)	9.19	13.88	17.69
DPPH(% Inhibition)	41.89	51.24	49.91
Total phenolic content Folin-Ciocalteu (mg GAE/g DM)	63.12	94.30	126.03

[1] CON: Control. [2] WSS5: Whole sesame seeds at 5%. [3] WSS10: Whole sesame seeds at 10%. [4] NDF: Neutral detergent fiber. [5] ADF: Acid detergent fiber [6] FRAP: Ferric reducing ability of plasma.

2.2. Feed Sample Analyses

Samples of the alfalfa hay, wheat straw and concentrate were analyzed for organic matter (OM; Official Method 7.009), dry matter (DM; Official Method 7.007) and crude protein (CP; Official Method 7.016) according to the AOAC (1984) and for neutral detergent fiber (NDF) and acid detergent fiber (ADF)-expressed exclusive of residual ash-according to the methods of Van Soest et al. [34].

2.3. *Blood Samples*

2.3.1. Blood Sample Collection for Neutrophil Isolation

Blood samples were taken at the 30th, 60th and 90th day from the beginning of the experiment for neutrophil isolation from the jugular vein into 17 Units/mL heparine-containing tubes (BD Vacutainer, Plymouth, UK).

2.3.2. Cell Isolation

Cell isolation was performed according to Tsiplakou et al. [35]. More specifically, isolation of neutrophils is carried out using density gradient centrifugation Histopaque 1077 (Sigma-Aldrich, St. Louis, MO, USA). Analytically, whole blood mixed with an equal volume of Hanks' balanced salt solution and three parts diluted blood layered on two parts Histopaque. Samples were centrifuged for 40 min at 500× *g*, 4 °C with the minimum acceleration and deceleration. After centrifugation, the upper phases were rejected. Neutrophils cells, which remained in the red cell layer, were lysed with the addition of endotoxin-free ultrapure water, and were vigorously shaken. NaCl was then added to resuspend cells in an isotonic solution (0.9% NaCl). These cells were washed several times and centrifuged for 5 min at 1000 g and 4 °C until a white and consistent cell pellet was clearly visible at the bottom of the tube. In the sequel, the final cell suspensions were cultured in 1 mL of growth medium RPMI (Sigma-Aldrich, St. Louis, MO, USA) which is incubated at 37 °C and then centrifuged at 1000 g for 5 min at 4 °C. Finally, the resulting cell pellets were again washed at least twice in 0.5 mL of phosphate-buffered saline (PBS) and centrifuged at 700 rpm for 1 min at 4 °C.

2.3.3. RNA Extraction

The isolated cells were homogenized with TRIzol™ (Invitrogen, Carlsbad, CA, USA) and after centrifugation with 24:23:1 phenol: chloroform: isoamyl alcohol solution, three distinct layers were obtained. The upper clear aqueous phase containing the RNA was transferred carefully into a new tube, without disturbing the interphase. RNA pellet was precipitated with 70% ethanol and then was dissolved in milli-Q water. The quantity and quality of the extracted RNA were evaluated by ND-1000 spectrophotometer (NanoDrop, Wilmington, DE, USA); the quantity was measured in ng/μL, and its purity was determined based on the A260/A280 and A260/A23 ratios. In addition, RNA integrity was assessed by electrophoresis on an agarose gel. As defined, the isolated RNA was treated with Turbo™ DNase I (2U/μL, commercially available kit: Invitrogen, Carlsbad, CA, USA), accordingly to the manufacturer's instructions. Absence of genomic DNA contamination was confirmed by PCR, using glycer aldehyde3-phosphatedehydro genase (GAPDH; housekeeping gene) Then, RNA samples were further purified by using phenol: chloroform and ethanol precipitation. The quantity and quality of the pure RNA samples were again confirmed by spectrophotometry (NanoDropND-1000) and by agarose gel (0.7%) electrophoresis.

2.3.4. cDNA Synthesis

Approximately 500 ng of RNA was used per cDNA synthesis by using the Prime Script First Strand cDNA Synthesis Kit (Takara, Shiga, Japan) according to the manufacturer's protocol using a mix of random hexamers and oligo-dT primers.

2.3.5. Primers

To derive primers sequences the ARS1 goat annotation was used. A pair of primers specific for each target gene (Table 2) were used by previous studies [35,36] designed to be specific for *Capra hircus* by using Primer Express Software (version 3.0) and verified using the Geneious Software (Biomatters, Auckland, New Zealand) and were tested against genomic DNAs to confirm that a single amplicon of 70 bp would result from quantitative real-time PCR (qPCR). In addition, dissociation curves were generated, and the amplification products were subjected to agarose gel electrophoresis to confirm the production of a single amplicon per reaction.

Table 2. Primers used for real-time qPCR and the mean PCR efficiency for each gene as calculated by LinRegPCR software [37].

Gene	Forward Primer 5′-3′	Reverse Primer 5′-3′	Ensemble
NLRC3	CAACCTACTCCACGACCAGG	TGGATGAAGTTCCACTGCA	ENSG00000167984
TLR4	ATGAACCACTCCACTCGCTC	TCTTGCTCCTTAGAGGCCGT	ENSG00000136869
NF-KB	AAGCTGTGGTGGAGGACTTG	ACAGAGTTACCCAAGCGGTC	ENSG00000109320
MYD88	ACAGACAAACTATCGGCTGA	CACCTCTTCTCAATGAGTTCA	ENSG00000172936
MAPK1	GCAACGACCACATCTGCTAC	AGGTTGGAAGGCTTGAGGTC	ENSG00000100030
IL1A	TCAAGCCCAGATCAGCACAT	TGATTGAGGGCGTCGTTCAG	ENSG00000115008
IL1B	TGGATAGCCCATGTGTGCTG	CAGAACACCACTTCTCGGCT	ENSG00000125538
TNFA	GGGAGACACAAACTAAGGGCT	AACCTGCAGTTCAGCTCCG	ENSG00000232810
TNFB	ACTCCCGAAGCCCTTCACCCG	GGCGGAGGAAGGCGCGGTCCG	ENSG00000226979
IL2	AAATCCCGAGAACCTCAAGCT	TGTAGCGTTAACCTTGGGCA	ENSG00000109471
IL6	CAGCAAGGAGACACTGGCAGA	TCCATCTTTTTCCTCCATTTTTGG	ENSG00000136244
STAT3	CGCAATTAGGCAGAGCAACTG	CCCTGTATCAGAGACCATCCCA	ENSG00000168610
TRIF	GCACGTCTAGCCTGCTTAC	TTGCGGGCCCGCAGCATCT	ENSG00000127666
IRF3	CCAGAGGCTGGGGCACTGCC	CCTTCGGGACCTCGCCGTCA	ENSG00000126456
IFNG	AAATTCCGGTGGATGATCTG	ACCATTACATTGATGCTCTCC	ENSG00000111537
TRAF3	TAACTGCTGCATTCGCTCCA	GGAACACAAAGCTGGGGTTG	ENSG00000131323
IRF5	ACATCCCCAGTGAGAAGCAG	ATGGCATACAGATCCTGGCC	ENSG00000128604
CCL5	CAAGTGCTCCATGGCAGCAG	GTTGGCGCACACCTGACG	ENSG00000271503
IL8	CCTGCTCTCTGCAGCTCTGTG	TGCATTGGCATCGAAGTTCTG	ENSG00000169429
CXCL16	GTGCCTGTGTTGTCCCTCTT	GCTTGCACACCACGTAGAGT	ENSG00000161921
HO1	GAGCTGACCCGAGAAGGTTT	AGACGGGGTTCTCCTTGTTG	ENSG00000100292
IL10	CTGGGGGAGAAGCTGAAGAC	CTCTCTTCACCTGCTCCACC	ENSG00000136634
JUND	ACGCAGTTCCTCTTTCCCAA	CCAGCTGGTTTTGCTTGTGT	ENSG00000130522
CHUK	TGCAGGGAAAGAGGCAGAAA	GACCGAGCAGAACTCTGTGT	ENSG00000213341
GAPDH	AAAGCCATCACCATCTTCCA	ACCACGTACTCAGCACCTCAT	ENSG00000111640
YWHAZ	TGTTCTATTGTGCCTAGTACACTGT	CATCAAGACTCACTGCCTCCC	ENSG00000164924

2.3.6. Real-Time Quantitative PCR

The expression levels of genes were estimated by a Step One Plus ™ Real-Time PCR System (Applied Biosystems, Foster City, CA, USA) using SYBR Select Master Mix (Applied Biosystems, Austin, TX, USA), gene-specific primers at a final concentration of 0.2 μM each (forward and reverse) and 1 μL of each cDNA as template. Thermal cycling was started with denaturation at 95 °C for 15 min, followed by 40 cycles at 95 °C for 15 s and 62 °C for 10 s. GAPDH and YWHAZ were used as housekeeping genes to normalize the cDNA template concentrations. The choice of housekeeping genes was based on a study by Vorachek et al. [38].

2.3.7. Normalization

The expression levels of the genes were calculated as $(1 + E)^{-\Delta Ct}$, where ΔCt is the difference between the geometric mean of the two housekeeping genes' Cts and the Ct of the target gene, and the primer efficiency is the mean of each amplicon's efficiency per primer, which was calculated by employing the linear regression method on the log (fluorescence) per cycle number (ΔRn) using the LinReg PCR software [37].

2.3.8. Statistical Analysis

Experimental data are presented as least squares means ± standard errors and were analyzed using a general linear model (GLM) for repeated measures, considering the sampling time (T) as the repeated measure, with fixed effects of dietary treatments (D) (CON, WSS5, WSS10), sampling time (T) (30th, 60th, 90th experimental day) and the interactions among them (D × T) according to the model:

$$Yijk = \mu + Di + Tj + (D \times T)\, ij + Ak + eijk \tag{1}$$

where ϒijk is the dependent variable, μ the overall mean, Di the effect of dietary treatment (I = 1, 2, 3), Tj the effect of sampling time (j = 1, 2, 3), (D × T) ij the interaction between dietary treatments and sampling time, Ak the animal's random effect and eijk the residual error. Post hoc analyses were performed when appropriate using Duncan's multiple range test. Kolmogorov-Smirnov test revealed that all variables followed a normal distribution. Pearson's correlation coefficients were used to determine the relationships between gene expression in neutrophils using heat map chart. For all tests, the significance was set at 0.05. Graphs were drawn using SPSS software (version 20.0, IBM, Armonk, NY, USA), and the error bars represent the standard error of the mean (SEM). Statistical analysis was performed using the statistical packages SPSS software (version 20.0, IBM, Armonk, NY, USA).

3. Results

A significant reduction in the expression levels of *MAPK1, IL6, TRIF, IFNG, TRAF3* and *JUND* genes in the neutrophils of WSS10 fed goats compared with the CON was found (Figure 1).

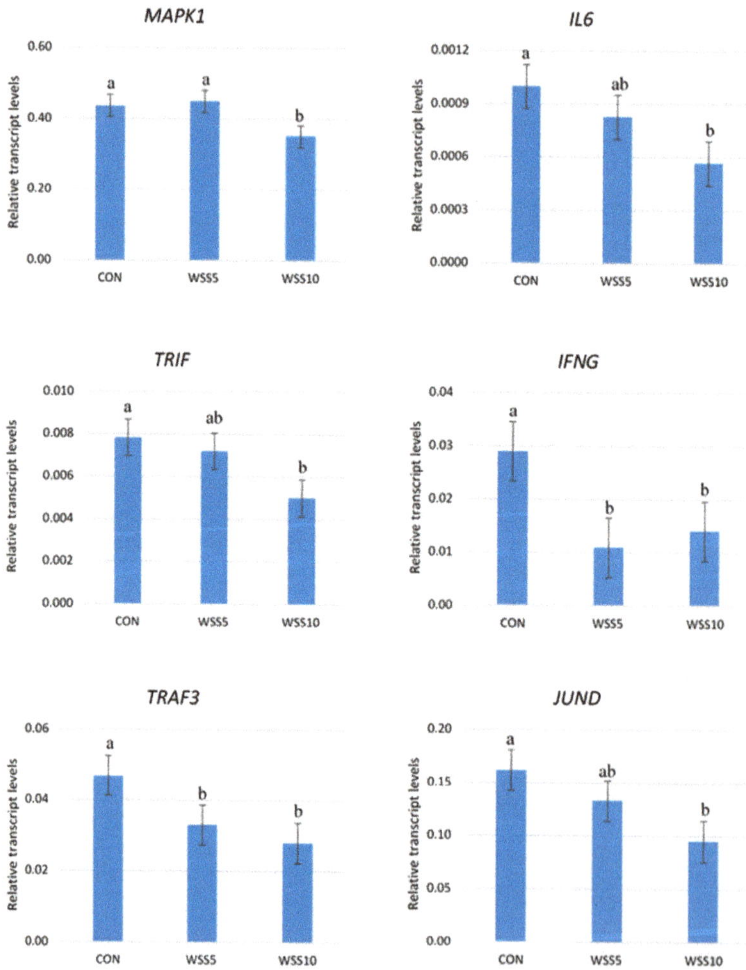

Figure 1. The Transcript abundance of several genes in the neutrophils of goats. Bars represent means ± SEM of each (n = 8) of the three dietary treatments; CON: control, basal diet; WSS5: basal diet + 5% whole sesame seed; WSS10: basal diet + 10% whole sesame seed in goats. For each gene, bars with different superscripts (a, b) between the three dietary treatments (CON, WSS5, WSS10) differ significantly ($p \leq 0.05$), according to the analysis of variance (ANOVA) using a general linear model (GLM) for repeated measures. Post hoc analysis was performed using Duncan's multiple range test.

The same trend was found for the IFNG and TRAF3 genes in the neutrophils of SS5 fed goats (Figure 1). No differences were found in the expression levels of the above genes between the treated groups (Supplementary Figure S1).

A significant reduction in the expression levels of *NF-KB*, *MYD88*, *MAPK1*, *TNFA*, and *STAT3* genes in the neutrophils of goats throughout the experimental period was observed (Supplementary Table S1). The opposite happened in the relative abundance transcripts of *IRF5* and *HO1* genes (Supplementary Table S1). The highest expression levels of *NLRC3*, *TNFB* and *TRIF* genes were indicated in the 60th experimental day while in this day the *TLR4* gene showed the lowest expression levels (Supplementary Table S1). A significant decline in the expression levels of *IL1A*, *IL-2* and *CCL5* genes was found in the 90th compared with the 30th and 60th experimental period, while the opposite trend was observed for the *IL10* and *CXXL-16* genes (Supplementary Table S1).

Significantly positive correlations between the expression levels of; *TLR4* with *NF-KB*, *MYD88*, and *IL1B*, *MYD88* with *NF-KB* and *IRF3*, *STAT3* with *MYD88* and *NF-KB*, *JUND* with *HO1*, *NF-KB* with *IRF3* genes respectively, as well as between *MAPK1* with *MYD88* and *TLR4* and *NLRC3* with *IL6* and *TNFB* genes respectively, were found (Figure 2).

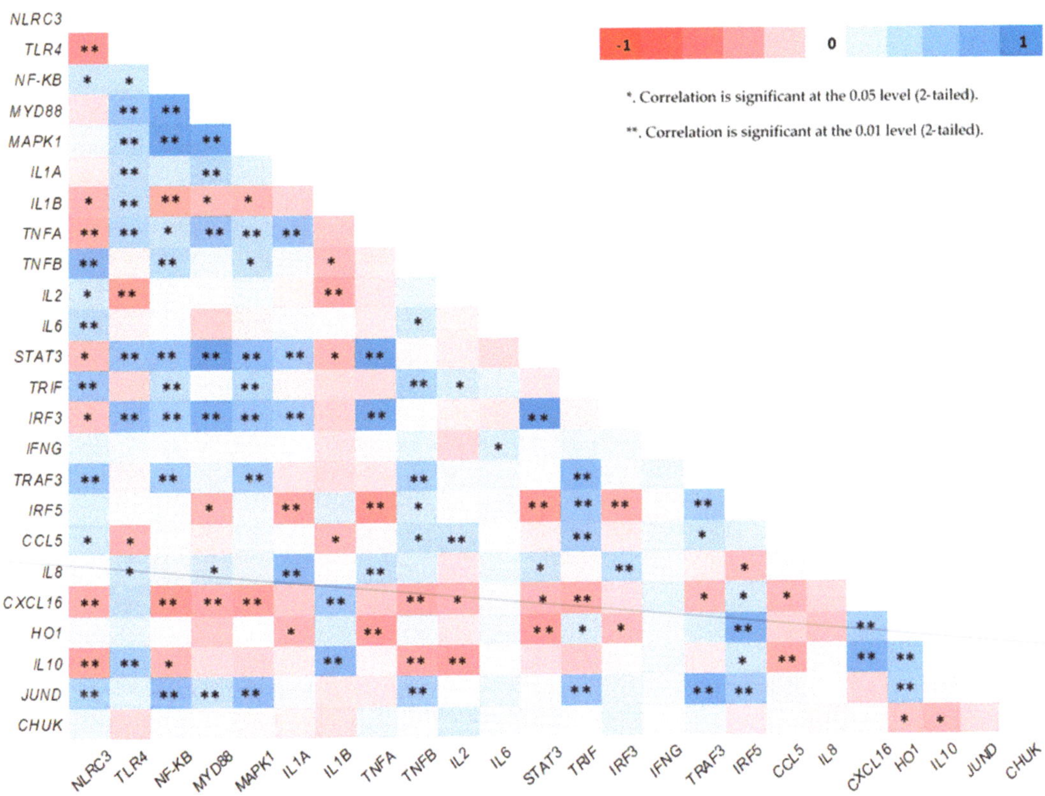

Figure 2. Pearson's heat map correlations between the expression level of several genes in neutrophils of goats.

4. Discussion

The impact of sesame seeds in the innate immunity of ruminants has rarely been investigated to the best of our knowledge. Neutrophils are involved in initiation of the inflammatory response [39], through the expression of several families of PRRs such as NLRs and TLRs [40,41] which can identify either microbial pathogens or components of host's cells that are released during cell damage or death. Significantly higher expression levels of *TLR4* gene have been found in blood neutrophils of ketotic cows [42]. Moreover, Roldan-Montes et al. [43] found in milk, a significant association between the identified polymorphisms of the *TLR4* gene and the somatic cell score of water buffaloes. A significant down-regulation in the expression of *TLR4* gene in LPS-stimulated BV-2 microglial cell line of rats was observed in vitro [14]. Thus, the results of this study referring to *NLRC3* and *TLR4* genes (Supplementary Figure S1) might show not only absence of any inflammation (clinical or subclinical) but also the protective role of sesame seeds in goats' cells survival.

TLR4 gene regulates the NF-KB pathway through the *MYD88* gene [44] which affects the MAPKs cascade [45]. Indeed, significantly positive correlations between the expression levels of *TLR4* with *NF-KB*, *TLR4* with *MYD88* and *MYD88* with *NF-KB* genes, were found (Figure 2). It was observed that sesamin, one of the main antioxidant compounds of sesame seeds, reduces the activation of NF-KB (measured by ELISA) and P38 MAPK kinase (measured by Western blot) in mice microglia cells treated with LPS [15]. The same has been shown for the expression of *TLR4* gene (measured by flow cytometry) in hepatic tissue of mice [46]. Sesamin inclusion in RPMI-8226 cells [47] and sesame oil aqueous extract in RAW 264.7 macrophages of mice treated with LPS [48] down regulate the expression of *NF-KB* gene in vitro. Additionally, a significant up-regulation in expression levels of *TLR4* and *MYD88* genes in goats' mammary epithelial cells [49], and in human endometrial cells [50] when stimulated in vitro by LPS have been found. Moreover, significantly higher expression levels of *MYD88* gene have been observed in bovine mastitis tissue [51]. Thus, the results of this study, concerning the *NF-KB*, *MYD88* and *MAPK1* genes, not only show no pathogens, stress or endogenous inflammatory factors in goats' organisms but also indicate a positive effect in their innate immunity when the animals were fed with the higher supplementation level (10%) of WSS.

NLRs and TLRs stimulate the MAPKs cascade thought the MYD88 pathway [52] and trigger the cytokines production [53,54]. The significantly positive correlations between the expression levels of *MAPK1* with *MYD88* and *TLR4* genes respectively, confirm this close relationship (Figure 2). Moreover, the positive correlations between the expression levels of *TLR4* and *IL1B* genes, as well as between the expression levels of *NLRC3* with *IL6* and *TNFB* genes respectively (Figure 2) show that both NLRs and TLRs regulate the cytokines expression. IL2 has anti-inflammatory properties [55] while IL6 is elevated in most cases of inflammation and have been recognized as target for therapeutic intervention [56]. Indeed, it has been found recently that elevated IL6 levels in blood plasma resulted a STAT3 hyperactivation in tumor cells [25]. However, sesamin has the ability to suppress the STAT3 signaling pathway (IL6/JAK/STAT3) in human hepatocellular carcinoma cell line HepG2 [57]. The anti-cancer effects of sesamin have been attributed to its ability to reduce significantly the expression of *NF-KB*, *IL6* and transcriptional target of *STAT3* [58]. In accordance with our findings, sesamin inhibits the expression levels of *IL6* gene, in a dose depend matter in vitro [15]. The anti-inflammatory activities of sesamin have been shown also, in influenza H1N1-induced peripheral blood mononuclear cells of humans by either the reduction in the expression levels of both *IL1B* and *TNFA* genes or the increase in the expression of *IL2* gene [59]. A significant reduction in the expression levels of IL2 gene was observed in cows infected with malignant catarrhal fever [60,61]. Furthermore, a significant down-regulation in the expression levels of pro-inflammatory (*IL1A, IL1B, TNFA*) genes including *IL6* was found in the liver of mice fed with a sesame oil rich diet [62]. Sesamin, reversed the inflammation which caused by the consumption of a high fat diet in rats by reducing the expression of *IL6* and *TNFA* genes [63]. Thus, the results of this study, as the *IL2, IL6, STAT3* and cytokines genes expression is concerned could high light as well the idea of an improvement of goats' innate immunity especially, when they were fed with the higher inclusion level of sesame seeds.

Although MYD88 is a common adaptor for all the TLRs except TLR3, TRIF, is an adaptor for TLR3 and TLR4 which promotes an alternative pathway that leads to the activation of IRF3 for induction of type IFN [19,64]. Our results referring to the *TRIF*, *IRF3* and *IFNG* genes, show that the highest dietary inclusion level (10%) of WSS affected also the MYD88 independent pathway. This is further supported by the changes in the expression levels of *TRAF* gene since both MYD88 and TRIF pathways are controlled by TRAF regulators such as TRAF3 [65,66]. Similar to our findings, a significant down-regulation in the expression levels of *IFNG* gene in cultured mononuclear cells of experimental autoimmune encephalomyelitis mice, fed with sesame oil, has been found [67]. Additionally, sesame oil reduces significantly the concentrations of IFNG in multiple sclerosis patients [68]. On the other hand, the expression level of *IRF3* gene increased significantly in goats' mammary epithelial cells after 3 h incubation in vitro with both toxins from LPS and gram-positive lipoteichoic acid bacterial [49]. The same was observed in bovine mammary epithelial cells when stimulated either with *Escherichia coli* or *Staphylococcus aureus* [69]. Furthermore, the pro-inflammatory role of IRF3 has been indicated also in mice macrophages, through the activation of TLR4-TRIF metabolic pathway which regulates the production of pro-inflammatory cytokines [70].

Chemokines such as CCL5, IL8 and CXCL16 can be produced by many cells including neutrophils [71] after proper stimulation [72]. So far, significant higher expression levels of *CCL5* gene have been observed in infected blood macrophages with *Mycobacterium* in vitro [73]. The same trend has been indicated in goats' mammary epithelial cells after incubation with gram-negative and/or gram-positive bacteria cell wall components in vitro [49]. The expression of *IL8* gene enhanced significantly in blood neutrophils of calving cows with clinical mastitis [71]. A positive correlation between *IL8* gene expression and the incidence of severe mastitis has been also shown [74]. However, chemokines such as IL8 can be also released from the cells as response to the reactive oxygen species (ROS) [75]. Additionally, CXCL16 chemokine can have a scavenger role for the uptake of oxidase molecules such as the low-density lipoproteins [76]. It has been shown that various antioxidant compounds can protect low-density lipoprotein (LDL) from oxidation in vitro [77]. Indeed, a delay in the oxidation of lipoproteins in the blood plasma of mice fed with sesame oil has been found due to its sesamin and sesamone content which was accompanied by a significant reduction in the CXCL16 blood plasma content [62]. Thus, the results of this study, concerning the expressions of chemokines (*CCL5*, *IL8* and *CXCL16*), further support the use of sesame seeds as a nutritional tool for the improvement of goats both innate immunity and antioxidant status.

HO1 is a highly inducible gene well known for its anti-inflammatory, immunomodulatory and antioxidants functions [78]. Similar with our findings, sesamin did not modify the expression levels of *HO1* gene in rats in vitro [79]. On the contrary, a significant up-regulation in the expression levels of *HO1* gene has been found in the liver of bovine and mice, infected by *Fachiola hepatica* [80]. The same was observed for both *HO1* and *IL10* genes in LPS-stimulated macrophages of mice [81]. Although the metabolic pathway which regulates *HO1* gene expression in not clear, activation of STAT3 by IL10 cytokine has repeatedly been suggested. The positive relationship between the expression levels of *HO1* and *IL10* genes ($p < 0.01$) supports this suggestion while the negative relation between the expression levels of *STAT3* and *IL10* genes ($p < 0.01$), which was found in this study (Figure 2), needs further investigation in order to clarify the role of *STAT3* gene in this metabolic pathway. Thus, referring to the results on the expression levels of *HO1*, *IL10* and *STAT3* genes of this study, an enhancement of the innate immune responses with the higher supplementation level of WSS could be claimed.

JUND gene might have also an involvement in the *HO1* gene expression. It has been found that JUND protein repressed *HO1* gene expression in human renal epithelial cells [82]. The relationship between the expression levels of *HO1* and *JUND* genes supports this link (Figure 2). Moreover, the expression levels of *JUND* gene followed the same trend with the expression level of *MAPK1* gene (Supplementary Table S1). *JUND* gene has a fundamental role in the defense against oxidative stress [83]. Thus, its sharpest down-regulation with the highest supplementation level of sesame seeds might show that the SS10 goats had a sufficient pool of antioxidants compounds in their organism such as sesamin, sesaminol, etc. which enhance their innate immunity.

So far, in the innate immunity, little attention has been given in inflammatory mediators such as the IkappaB kinase (IKK). The role of IKK-α subunit (*CHUK*) in inflammation is not well known. However, CHUK gene is required for the activation of the "alternative" NF-KB pathway which is activated by the TNF family cytokines [84]. Moreover, CHUK has anti-inflammatory role through the regulation of SUMO (small ubiquitin-related modifier) ligase activity of protein inhibitor of activated STAT1 (PIAS) [85]. In accordance with our findings sesamin had no effect on the expression of *CHUK* gene in various human cells lines in vitro [47]. More research is needed in order to clarify the role of *CHUK* gene in the innate immunity.

5. Conclusions

Overall, our study provides new evidence regarding the impact of dietary supplementation with WSS in the innate immunity of dairy goats. The highest inclusion level (WSS10) seems the best modulator of goats' innate immunity, as demonstrated by the sharpest decline in the expression levels of genes (*MAPK1*, *IL6*, *TRIF*, *IFNG*, *TRAF3*, and *JUND*) involved with inflammatory metabolic pathways. The topmost intake of WSS also regulates both MYD88 dependent (*MAPK1*) and independent (*TRIF*, *TRAF3*, *IFNG*) pathway, while this of WSS5 the independent one only. The above findings are very important in animal husbandry since inflammation should be limited as much as possible, and animals' innate immunity should be activated only when is needed in order to be stronger and more effective. Finally, lignans can eliminate the pro-inflammatory compound, which is produced by LA's metabolism, making WSS one of the best way to administer LA in goats' diet.

Supplementary Materials: The following are available online at https://www.mdpi.com/2076-2615/11/2/468/s1, Figure S1: The Transcript abundance of several genes in the neutrophils of goats. Bars represent means ± SEM of each (n = 8) of the three dietary treatments; CON: control, basal diet; WSS5: basal diet + 5% whole sesame seed; WSS10: basal diet + 10% whole sesame seed in goats. The analysis of variance (ANOVA) using a general linear model (GLM) for repeated measures revealed that for these genes there was not significant difference between the three dietary treatments ($p > 0.05$), Table S1: Transcript abundance of several genes in the neutrophils of goats: NOD-like receptor (*NLRC3*), Toll-like receptors 4 (*TLR4*), Nuclear factor kappa B (*NF-KB*), Myeloid-Differentiation-primary response gene 88 (MYD88), Mitogen-Activated Protein Kinase-1 (*MAPK1*), Interleukin 1 Alpha (*IL1A*), Interleukin 1 Beta (*IL1B*), Tumor necrosis factor Alpha (*TNFA*), Tumor necrosis factor Beta (*TNFB*), Interleukin 2 (*IL2*), Interleukin 6 (*IL6*), Signal Transducer and Activator of Transcription 3 (*STAT3*), TIR (Toll/Interleukin-1 Receptor) domain-containing adaptor protein inducing interferon beta (*TRIF*), Interferon Regulatory Factor 3 (*IRF3*), Interferon gamma (*IFNG*), TNF Receptor-associated Factor 3 (*TRAF3*), Interferon Regulatory Factor 5 (*IRF5*), C-C motif chemokine ligand 5 (*CCL5*), Interleukin 8 (*IL8*), Chemokine (C-X-C motif) ligand 16 (*CXCL16*), Heme Oxygenase-1 (*HO1*), Interleukin 10 (*IL10*), Transcription factor JunD (*JUND*) and Conserved Helix-Loop-Helix-Ubiquitous Kinase (*CHUK*) or IKKA relative to the geometrical mean of the references genes (Glyceraldehyde 3-Phosphate Dehydrogenase (*GAPDH*) and Tyrosine 3-monoxygenase/tryptophan 5-monooxygenase activation protein, zeta polypeptide (*YWHAZ*)).

Author Contributions: Conceptualization, E.T.; methodology, E.T., E.F. and C.M.; software, K.S and D.S.; validation, E.T., K.S., E.F., D.S. and C.M.; formal analysis, C.M.; investigation, E.T.; data curation, E.T., K.S. and C.M.; writing—original draft preparation, E.T. and C.M.; writing—review and editing, E.T. and E.F.; visualization, C.M.; supervision, E.T.; project administration, E.T. All authors have read and agreed to the published version of the manuscript.

Funding: This research is co-financed by Greece and the European Union (European Social Fund-ESF) through the Operational Programme «Human Resources Development, Education and Lifelong Learning» in the context of the project "Strengthening Human Resources Research Potential via Doctorate Research" (MIS-5000432), implemented by the State Scholarships Foundation (IKY)».

Institutional Review Board Statement: The study was conducted according to the guidelines of Ethical Committee guidelines of Faculty of Animal Science of Agricultural University of Athens (026/10022017).

Data Availability Statement: Not applicable.

Conflicts of Interest: The authors declare no conflict of interest.

References

1. Calder, P.C.; Bosco, N.; Bourdet-Sicard, R.; Capuron, L.; Delzenne, N.; Doré, J.; Franceschi, C.; Lehtinen, M.J.; Recker, T.; Salvioli, S.; et al. Health relevance of the modification of low grade inflammation in ageing (inflammageing) and the role of nutrition. *Ageing Res. Rev.* **2017**, *40*, 95–119. [CrossRef] [PubMed]
2. Bannenberg, G.; Serhan, C.N. Specialized pro-resolving lipid mediators in the inflammatory response: An update. *Biochim. Biophys. Acta.* **2010**, *1801*, 1260–1273. [CrossRef]
3. Senftleber, N.K.; Nielsen, S.M.; Andersen, J.R.; Bliddal, H.; Tarp, S.; Lauritzen, L.; Furst, D.E.; Suarez-Almazor, M.E.; Lyddiatt, A.; Christensen, R. Marine Oil Supplements for Arthritis Pain: A Systematic Review and Meta-Analysis of Randomized Trials. *Nutrients* **2017**, *9*, 42. [CrossRef] [PubMed]
4. Ramsden, C.E.; Ringel, A.; Feldstein, A.E.; Taha, A.Y.; MacIntosh, B.A.; Hibbeln, J.R.; Majchrzak-Hong, S.F.; Faurot, K.R.; Rapoport, S.I.; Cheon, Y.; et al. Lowering dietary linoleic acid reduces bioactive oxidized linoleic acid metabolites in humans. *Prostaglandins Leukot. Essent Fatty Acids* **2012**, *87*, 135–141. [CrossRef]
5. Ilich, J.Z.; Kelly, O.J.; Kim, Y.; Spicer, M.T. Low-grade chronic inflammation perpetuated by modern diet as a promoter of obesity and osteoporosis. *Arh. Hig. Rada Toksikol.* **2014**, *65*, 139–148. [CrossRef]
6. Shrestha, N.; Cuffe, J.S.; Holland, O.J.; Bulmer, A.C.; Hill, M.; Perkins, A.V.; Muhlhausler, B.S.; McAinch, A.J.; Hryciw, D.H. Elevated maternal linoleic acid reduces circulating leptin concentrations, cholesterol levels and male fetal survival in rat model. *J. Physiol.* **2019**, *597*, 3349–3361. [CrossRef] [PubMed]
7. Marchix, J.; Choque, B.; Kouba, M.; Fautrel, A.; Catheline, D.; Legrand, P. Excessive dietary linoleic acid induces proinflammatory markers in rats. *J. Nutr. Biochem.* **2015**, *12*, 1434–1441. [CrossRef] [PubMed]
8. Froyen, E.; Burns-Whitmore, B. The Effects of Linoleic Acid Consumption on Lipid Risk Markers for Cardiovascular Disease in Healthy Individuals: A Review of Human Intervention Trials. *Nutrients* **2020**, *12*, 2329. [CrossRef]
9. Chowdhury, R.; Warnakula, S.; Kunutsor, S.; Crowe, F.; Ward, H.A.; Johnson, L.; Franco, O.H.; Butterworth, A.S.; Forouhi, N.G.; Thompson, S.G.; et al. Association of dietary, circulating, and supplement fatty acids with coronary risk: A systematic review and meta-analysis. *Ann. Intern. Med.* **2014**, *160*, 398–406. [CrossRef]
10. Hooper, L.; Al-Khudairy, L.; Abdelhamid, A.S.; Rees, K.; Brainard, J.S.; Brown, T.J.; Ajabnoor, S.M.; O'Brien, A.T.; Winstanley, L.E.; Donaldson, D.H.; et al. Omega-6 fats for the primary and secondary prevention of cardiovascular disease. *Cochrane Database Syst. Rev.* **2018**, *7*. [CrossRef]
11. Hansen, R. Sesame Profile. Available online: http://www.agmrc.org/commodities__products/grains__oilseeds/sesame_profile (accessed on 19 August 2011).
12. Khorrami, S.; Daneshmandi, S.; Mosayeb, G. Sesame seeds essential oil and Sesamol modulate the pro-inflammatory function of macrophages and dendritic cells and promote Th2 response. *Med. J. Islam Repub. Iran.* **2018**, *32*, 566–573. [CrossRef]
13. Wu, M.S.; Aquino, L.; Barbaza, M.; Hsieh, C.L.; Castro-Cruz, K.A.; Yang, L.L.; Tsai, P.W. Anti-Inflammatory and Anticancer Properties of Bioactive Compounds from Sesamum indicum L.-A Review. *Molecules* **2019**, *24*, 4426. [CrossRef]
14. Udomruk, S.; Kaewmool, C.; Pothacharoen, P.; Phitak, T.; Kongtawelert, P. Sesamin suppresses LPS-induced microglial activation via regulation of TLR4 expression. *J. Funct. Foods* **2018**, *49*, 32–43. [CrossRef]
15. Jeng, K.C.; Hou, R.C.; Wang, J.C.; Ping, L.I. Sesamin inhibits lipopolysaccharide-induced cytokine production by suppression of p38 mitogen-activated protein kinase and nuclear factor-κB. *Immunol. Lett.* **2005**, *97*, 101–106. [CrossRef] [PubMed]
16. Katayama, S.; Sugiyama, H.; Kushimoto, S.; Uchiyama, Y.; Hirano, M.; Nakamura, S. Effects of Sesaminol Feeding on Brain Aβ Accumulation in a Senescence-Accelerated Mouse-Prone 8. *J. Agric. Food Chem.* **2016**, *64*, 4908–4913. [CrossRef] [PubMed]
17. Messaoudi, I.; Estep, R.; Robinson, B.; Wong, S.W. Nonhuman Primate Models of Human Immunology. *Antioxid. Redox Signal.* **2011**, *14*, 261–273. [CrossRef]
18. Rosales, C.; Demaurex, N.; Lowell, C.A.; Uribe-Querol, E. Neutrophils: Their Role in Innate and Adaptive Immunity. *J. Immunol. Res.* **2016**, *2016*, 1469780. [CrossRef]
19. Kawasaki, T.; Kawai, T. Toll-Like Receptor Signaling Pathways. *Front. Immunol.* **2014**, *5*. [CrossRef] [PubMed]
20. Majewska, M.; Szczepanik, M. The role of Toll-like receptors (TLR) in innate and adaptive immune responses and their function in immune response regulation. *Postepy. Hig. Med. Dosw.* **2006**, *60*, 52–63.
21. Huang, W.; Hung, M. Beyond NF-κB activation: Nuclear functions of IκB kinase α. *J. Biomed. Sci.* **2013**, *20*. [CrossRef]
22. Roman-Blas, J.A.; Jimenez, S.A. NF-kappaB as a potential therapeutic target in osteoarthritis and rheumatoid arthritis. *Osteoarthr. Cartil.* **2006**, *14*, 839–848. [CrossRef]
23. Cho, J.W.; Lee, K.S.; Kim, C.W. Curcumin attenuates the expression of IL-1beta, IL-6, and TNF-alpha as well as cyclin E in TNF-alpha-treated HaCaT cells; NF-κB and MAPKs as potential upstream targets. *Int. J. Mol. Med.* **2007**, *19*, 469–474. [CrossRef]

24. Wong, S.W.; Kwon, M.J.; Choi, A.M.; Kim, H.P.; Nakahira, K.; Hwang, D.H. Fatty acids modulate toll-like receptor 4 activation through regulation of receptor dimerization and recruitment into lipid rafts in a reactive oxygen species-dependent manner. *J. Biol. Chem.* **2009**, *284*, 27384–27392. [CrossRef]
25. Johnson, D.E.; O'Keefe, R.A.; Grandis, J.R. Targeting the IL-6/JAK/STAT3 signalling axis in cancer. *Nat. Rev. Clin. Oncol.* **2018**, *15*, 234–248. [CrossRef]
26. Medzhitov, R. Toll-like receptors and innate immunity. *Nat. Rev. Immunol.* **2001**, *1*, 135–145. [CrossRef]
27. Takeda, K.; Akira, S. Toll-like receptors in innate immunity. *Int. Immunol.* **2005**, *17*, 1–14. [CrossRef] [PubMed]
28. Takaoka, A.; Yanai, H.; Kondo, S.; Duncan, G.; Negishi, H.; Mizutani, T.; Kano, S.-I.; Honda, K.; Ohba, Y.; Mak, T.W.; et al. Integral role of IRF-5 in the gene induction programme activated by Toll-like receptors. *Nature* **2005**, *434*, 243–249. [CrossRef] [PubMed]
29. Barton, G.M.; Kagan, J.C. A cell biological view of Toll-like receptor function: Regulation through compartmentalization. *Nat. Rev. Immunol.* **2009**, *9*, 535–542. [CrossRef]
30. Xie, X.; Jin, J.; Zhu, L.; Jie, Z.; Li, Y.; Zhao, B.; Cheng, X.; Li, P.; Sun, S.-C. Cell type-specific function of TRAF2 and TRAF3 in regulating type I IFN induction. *Cell Biosci.* **2019**, *9*. [CrossRef] [PubMed]
31. Hernandez, J.M.; Floyd, D.H.; Weilbaecher, K.N.; Green, P.L.; Boris-Lawrie, K. Multiple facets of junD gene expression are atypical among AP-1 family members. *Oncogene* **2008**, *27*, 4757–4767. [CrossRef]
32. Schumacher, A.; Zenclussen, A.C. Effects of heme oxygenase-1 on innate and adaptive immune responses promoting pregnancy success and allograft tolerance. *Front. Pharmacol.* **2015**, *5*, 288. [CrossRef]
33. National Research Council. *Nutrient Requirements of Small Ruminants: Sheep, Goats, Cervids, and New World Camelids*; The National Academies Press: Washington, DC, USA, 2007. [CrossRef]
34. Van Soest, P.J.; Robertson, J.B.; Lewis, B.A. Methods for dietary fiber, neutral detergent fiber, and nonstarch polysaccharides in relation to animal nutrition. *J. Dairy Sci.* **1991**, *74*, 3583–3597. [CrossRef]
35. Tsiplakou, E.; Mavrommatis, A.; Skliros, D.; Sotirakoglou, K.; Flemetakis, E.; Zervas, G. The effects of dietary supplementation with rumen-protected amino acids on the expression of several genes involved in the immune system of dairy sheep. *J. Anim. Physiol. Anim. Nutr.* **2018**, *102*, 1437–1449. [CrossRef]
36. Tsiplakou, E.; Mavrommatis, A.; Skliros, D.; Righi, F.; Flemetakis, E. The impact of rumen-protected amino acids on the expression of key- genes involved in the innate immunity of dairy sheep. *PLoS ONE* **2020**, *15*, e0233192. [CrossRef] [PubMed]
37. Ramakers, C.; Ruijter, J.M.; Deprez, R.H.; Moorman, A.F. Assumption-free analysis of quantitative real-time polymerase chain reaction (PCR) data. *Neurosci. Lett.* **2003**, *339*, 62–66. [CrossRef]
38. Vorachek, W.; Hugejiletu Bobe, G.; Hall, J. Reference gene selection for quantitative PCR studies in sheep neutrophils. *Int. J. Mol. Sci.* **2013**, *14*, 11484–11495. [CrossRef]
39. Rosales, C.; Lowell, C.A.; Schnoor, M.; Uribe-Querol, E. Neutrophils: Their Role in Innate and Adaptive Immunity. *J. Immunol. Res.* **2017**, *2017*. [CrossRef]
40. Wu, B.; Peisley, A.; Tetrault, D. Molecular imprinting as a signal activation mechanism of the viral RNA sensor RIG-I. *Mol. Cell* **2014**, *55*, 511–523. [CrossRef] [PubMed]
41. Hu, H.; Sun, S.C. Ubiquitin signaling immune responses. *Cell Res.* **2016**, *26*, 457–483. [CrossRef]
42. Zhang, Y.; Li, X.; Zhang, H.; Zhao, Z.; Peng, Z.; Wang, Z.; Liu, G.; Li, X. Non-Esterified Fatty Acids Over-Activate the TLR2/4-NF-Kb Signaling Pathway to Increase Inflammatory Cytokine Synthesis in Neutrophils from Ketotic Cows. *Cell Physiol. Biochem.* **2018**, *48*, 827–837. [CrossRef]
43. Roldan-Montes, V.; Cardoso, D.F.; Hurtado-Lugo, N.A.; do Nascimento, A.V.; Santos, D.; Scalez, D.; de Freitas, A.C.; Herrera, A.C.; Albuquerque, L.G.; de Camargo, G.; et al. Polymorphisms in TLR4 Gene Associated with Somatic Cell Score in Water Buffaloes (*Bubalus bubalis*). *Front. Vet. Sci.* **2020**, *7*, 568249. [CrossRef] [PubMed]
44. Calder, P.C. Fatty acids Long-chain fatty acids and inflammation. *Proc. Nutr. Soc.* **2012**, *71*, 274–289. [CrossRef]
45. Akira, S.; Uematsu, S.; Takeuchi, O. Pathogen recognition and innate immunity. *Cell* **2006**, *124*, 783–801. [CrossRef]
46. Ma, L.; Gong, X.; Kuang, G.; Jiang, R.; Chen, R.; Wan, J. Sesamin ameliorates lipopolysaccharide/d-galactosamine-induced fulminant hepatic failure by suppression of Toll-like receptor 4 signaling in mice. *Biochem. Biophys. Res. Commun.* **2015**, *461*, 230–236. [CrossRef]
47. Harikumar, K.B.; Sung, B.; Tharakan, S.T.; Pandey, M.K.; Joy, B.; Guha, S.; Krishnan, S.; Aggarwal, B.B. Sesamin manifests chemopreventive effects through the suppression of NF-kappa B-regulated cell survival, proliferation, invasion, and angiogenic gene products. *Mol. Cancer Res.* **2010**, *8*, 751–761. [CrossRef]
48. Selvarajan, K.; Narasimhulu, C.A.; Bapputty, R.; Parthasarathy, S. Anti-inflammatory and antioxidant activities of the nonlipid (aqueous) components of sesame oil: Potential use in atherosclerosis. *J. Med. Food* **2015**, *18*, 393–402. [CrossRef]
49. Bulgari, O.; Dong, X.; Roca, A.L.; Caroli, A.M.; Loor, J.J. Innate immune responses induced by lipopolysaccharide and lipoteichoic acid in primary goat mammary epithelial cells. *J. Anim. Sci. Biotechnol.* **2017**, *8*, 29–38. [CrossRef]
50. Rashidi, N.; Mirahmadian, M.; Jeddi-Tehrani, M.; Rezania, S.; Ghasemi, J.; Kazemnejad, S. Lipopolysaccharide and lipoteichoic acid-mediated pro-inflammatory cytokine production and modulation of TLR2, TLR4 and MyD88 expression in human endometrial cells. *J. Reprod. Infertil.* **2005**, *16*, 72–81.
51. Wu, Z.; Li, F.; Liu, D.; Xue, H.; Zhao, X. Novel Type XII Staphylococcal cassette chromosome mec harboring a new cassette chromosome recombinase CcrC2. *Antimicrob. Agents Chemother.* **2015**, *59*, 7597–7601. [CrossRef]

52. Qian, C.; Cao, X. Regulation of Toll-like receptor signaling pathways in innate immune responses. *Ann. N. Y. Acad. Sci.* **2012**, *1283*, 67–74. [CrossRef] [PubMed]
53. Zhang, Y.L.; Dong, C. MAP kinases in immune responses. *Cell Mol. Immunol.* **2005**, *2*, 20–27.
54. Lim, M.X.; Png, C.W.; Tay, C.Y.; Teo, J.D.; Jiao, H.; Lehming, N.; Tan, K.S.; Zhang, Y. Differential regulation of proinflammatory cytokine expression by mitogen-activated protein kinases in macrophages in response to intestinal parasite infection. *Infect. Immun.* **2014**, *82*, 4789–4801. [CrossRef]
55. Bachmann, M.F.; Wolint, P.; Walton, S.; Schwarz, K.; Oxenius, A. Differential role of IL-2R signaling for CD8+ T cell responses in acute and chronic viral infections. *Eur. J. Immunol.* **2007**, *37*, 1502–1512. [CrossRef]
56. Scheller, J.; Chalaris, A.; Schmidt-Arras, D.; Rose-John, S. The pro-and anti-inflammatory properties of the cytokine inteleukin-6. *Biochim. Biophys. Acta* **2011**, *1813*, 878–888. [CrossRef]
57. Deng, P.; Wang, C.; Chen, L.; Wang, C.; Du, Y.; Yan, X.; Chen, M.; Yang, G.; He, G. Sesamin induces cell cycle arrest and apoptosis through the inhibition of signal transducer and activator of transcription 3 signalling in human hepatocellular carcinoma cell line HepG2. *Biol. Pharm. Bull.* **2013**, *36*, 1540–1548. [CrossRef]
58. Kong, X.; Ma, M.Z.; Zhang, Y.; Weng, M.Z.; Gong, W.; Guo, L.Q.; Zhang, J.X.; Wang, G.D.; Su, Q.; Quan, Z.W.; et al. Differentiation therapy: Sesamin as an effective agent in targeting cancer stem-like side population cells of human gallbladder carcinoma. *BMC Complement. Altern Med.* **2014**, *14*, 254. [CrossRef]
59. Fanhchaksai, K.; Kodchakorn, K.; Pothacharoen, P.; Kongtawelert, P. Effect of sesamin against cytokine production from influenza type A H1N1-induced peripheral blood mononuclear cells: Computational and experimental studies. *In Vitro Cell Dev. Biol. Anim.* **2015**, *52*, 107–119. [CrossRef]
60. Meier-Trummer, C.S.; Rehrauer, H.; Franchini, M.; Patrignani, A.; Wagner, U.; Ackermann, M. Malignant catarrhal fever of cattle is associated with low abundance of IL-2 transcript and a predominantly latent profile of ovine Herpesvirus 2 gene expression. *PLoS ONE* **2009**, *4*, 6265. [CrossRef] [PubMed]
61. Russell, G.C.; Benavides, J.; Grant, D.M.; Todd, H.; Thomson, J.; Puri, V.; Nath, M.; Haig, D.M. Host gene expression changes in cattle infected with Alcelaphine herpesvirus 1. *Virus Res.* **2012**, *169*, 246–254. [CrossRef] [PubMed]
62. Narasimhulu, C.A.; Selvarajan, K.; Litvinov, D.; Parthasarathy, S. Anti-atherosclerotic and anti-inflammatory actions of sesame oil. *J. Med. Food* **2015**, *18*, 11–20. [CrossRef] [PubMed]
63. Zhang, R.; Yu, Y.; Hu, S.; Zhang, J.; Yang, H.; Han, B.; Cheng, Y.; Luo, X. Sesamin ameliorates hepatic steatosis and inflammation in rats on a high-fat diet via LXRα and PPARα. *Nutr. Res.* **2016**, *36*, 1022–1030. [CrossRef] [PubMed]
64. Cao, D.; Luo, J.; Chen, D.; Xu, H.; Shi, H.; Jing, X.; Zang, W. CD36 regulates lipopolysaccharide-induced signalling pathways and mediates the internalization of Escherichia coli in cooperation with TLR4 in goat mammary gland epithelial cells. *Sci. Rep.* **2016**, *6*, 23132. [CrossRef] [PubMed]
65. Hacker, H.; Tseng, P.H.; Karin, M. Expanding TRAF function: TRAF3 as a tri-faced immune regulator. *Nat. Rev. Immunol.* **2011**, *7*, 457–468. [CrossRef]
66. Yang, X.D.; Sun, S.C. Targeting signaling factors for degradation an emerging mechanism for TRAF functions. *Immunol. Rev.* **2015**, *266*, 56–71. [CrossRef] [PubMed]
67. Javan, M.R.; Zamani, M.R.; Aslani, S.; Dargahi Abbasabad, G.; Beirami Khalaj, M.; Serati-Nouri, H. Cytokine Modulatory Effects of Sesamum Indicum Seeds Oil Ameliorate Mice with Experimental Autoimmune Encephalomyelitis. *Arch. Asthma Allergy Immunol.* **2017**, *1*, 86–93. [CrossRef]
68. Faraji, F.; Hashemi, M.; Ghiasabadi, A.; Davoudian, S.; Talaie, A.; Ganji, A.; Mosayebi, G. Combination therapy with interferon beta-1a and sesame oil in multiple sclerosis. *Complement. Ther. Med.* **2019**, *45*, 275–279. [CrossRef]
69. Gilbert, F.B.; Cunha, P.; Jensen, K.; Glass, E.J.; Foucras, G.; Robert-Granie, C.; Rupp, R.; Rainard, P. Differential response of bovine mammary epithelial cells to Staphylococcus aureus or Escherichia coli agonists of the innate immune system. *Vet. Res.* **2013**, *44*, 40. [CrossRef] [PubMed]
70. Zhao, A.; Urban, J.F.; Anthony, R.M.; Sun, R.; Stiltz, J.; van Rooijen, N.; Wynn, T.A.; Gause, W.C.; Shea-Donohue, T. Th2 Cytokine-Induced Alterations in Intestinal Smooth Muscle Function Depend on Alternatively Activated Macrophages. *Gastroenterology* **2008**, *135*, 217–225. [CrossRef]
71. Alhussien, M.; Manjari, P.; Sheikh, A.A.; Mohammed Seman, M.; Reddi, S.; Mohanty, A.K.; Dang, A.K. Immunological attributes of blood and milk neutrophils isolated from crossbred cows during different physiological conditions. *Czech J. Anim. Sci.* **2016**, *61*, 223–231. [CrossRef]
72. Jarczak, J.; Kaba, J.; Reczyńska, D.; Bagnicka, E. Impaired expression of cytokines as a result of viral infections with an emphasis on small ruminant lentivirus infection in goats. *Viruses* **2016**, *8*, 186. [CrossRef] [PubMed]
73. Machugh, D.E.; Taraktsoglou, M.; Killick, K.E.; Nalpas, N.C.; Browne, J.A.; De Park, S.; Magee, D.A. Pan-genomic analysis of bovine monocyte-derived macrophage gene expression in response to in vitro infection with Mycobacterium avium subspecies paratuberculosis. *Vet. Res.* **2012**, *43*, 25. [CrossRef] [PubMed]
74. Galvao, K.N.; Pighetti, G.M.; Cheong, S.H.; Nydam, D.V.; Gilbert, R.O. Association between interleukin-8 receptor-α (CXCR1) polymorphism and disease incidence, production, reproduction, and survival in Holstein cows. *J. Dairy Sci.* **2011**, *94*, 2083–2091. [CrossRef] [PubMed]
75. DeForge, L.E.; Preston, A.M.; Takeuchi, E.; Kenney, J.; Boxer, L.A.; Remick, D.G. Regulation of interleukin 8 gene expression by oxidant stress. *J. Biol. Res.* **1993**, *268*, 25568–25576.

76. Lehrke, M.; Millington, S.C.; Lefterova, M.; Cumaranatunge, R.G.; Szapary, P.; Wilensky, R.; Rader, D.J.; Lazar, M.A.; Reilly, M.P. CXCL16 Is a Marker of Inflammation, Atherosclerosis, and Acute Coronary Syndromes in Humans. *J. Am. Coll. Cardiol.* **2007**, *49*, 442–449. [CrossRef]
77. Shariat, S.Z.A.S.; Mostafavi, S.A.; Khakpour, F. Antioxidant effects of vitamins c and e on the low-density lipoprotein oxidation mediated by myeloperoxidase. *Iran. Biomed. J.* **2013**, *17*, 22–28. [CrossRef]
78. Paine, A.; Eiz-Vesper, B.; Blasczyk, R.; Immenschuh, S. Signaling to heme oxygenase-1 and its anti-inflammatory therapeutic potential. *Biochem. Pharmacol.* **2010**, *80*, 1895–1903. [CrossRef] [PubMed]
79. Fukunaga, M.; Ohnishi, M.; Shiratsuchi, A.; Kawakami, T.; Takahashi, M.; Motomura, M.; Egusa, K.; Urasaki, T.; Inoue, A. Sesamin increases heme oxygenase-1 protein in RAW 264.7 macrophages through inhibiting its ubiquitination process. *Eur. J. Pharmacol.* **2014**, *741*, 214–221. [CrossRef]
80. Carasi, P.; Racedo, S.M.; Jacquot, C.; Elie, A.M.; Serradell, M.L.; Urdaci, M.C. Enterococcus durans EP1 a Promising Anti-inflammatory Probiotic Able to Stimulate sIgA and to Increase *Faecalibacterium prausnitzii* Abundance. *Front. Immunol.* **2017**, *8*. [CrossRef] [PubMed]
81. Lee, T.-S.; Chau, L.-Y. Heme oxygenase-1 mediates the anti-inflammatory effect of interleukin-10 in mice. *Nat. Med.* **2002**, *8*, 240–246. [CrossRef]
82. Hock, T.D.; Liby, K.; Wright, M.M.; McConnell, S.; Schorpp-Kistner, M.; Ryan, T.M.; Agarwal, A. JunB and JunD regulate human heme oxygenase-1 gene expression in renal epithelial cells. *J. Biol. Chem.* **2007**, *282*, 6875–6886. [CrossRef]
83. Raghunath, A.; Nagarajan, R.; Sundarraj, K.; Panneerselvam, L.; Perumal, E. Genome-wide identification and analysis of Nrf2 binding sites—Antioxidant response elements in zebrafish. *Toxicol. Appl. Pharmacol.* **2018**, *360*, 236–248. [CrossRef] [PubMed]
84. Lawrence, T. The Nuclear Factor NF- B Pathway in Inflammation. *Cold Spring Harbor. Perspect. Biol.* **2009**, *1*. [CrossRef] [PubMed]
85. Liu, B.; Yang, Y.; Chernishof, V.; Loo, R.R.O.; Jang, H.; Tahk, S.; Yang, R.; Mink, S.; Shultz, D.; Bellone, C.J.; et al. Proinflammatory Stimuli Induce IKKα-Mediated Phosphorylation of PIAS1 to Restrict Inflammation and Immunity. *Cell* **2007**, *129*, 903–914. [CrossRef] [PubMed]

animals

MDPI

Review

Application of Olive By-Products in Livestock with Emphasis on Small Ruminants: Implications on Rumen Function, Growth Performance, Milk and Meat Quality

Ouranios Tzamaloukas [1], Marina C. Neofytou [1] and Panagiotis E. Simitzis [2,*]

1 Department of Agricultural Sciences, Biotechnology and Food Science, Cyprus University of Technology, Limassol 3036, Cyprus; ouranios.tzamaloukas@cut.ac.cy (O.T.); maneofyt@gmail.com (M.C.N.)
2 Laboratory of Animal Breeding and Husbandry, Department of Animal Science, Agricultural University of Athens, Iera Odos, 11855 Athens, Greece
* Correspondence: pansimitzis@aua.gr; Tel.: +30-2105-294-427

Simple Summary: Olive oil is one of the main components in the Mediterranean diet that is known worldwide for its beneficial effects on human health due to its high content of monounsaturated fatty acids (MUFAs). During its extraction, a great quantity of olive by-products (OB) is generated that poses a risk to the environment due to its high organic load. Utilization of OB as a part of the ruminants' diet could minimize the costs related to animal feeding and OB management and contribute to the preservation of natural resources. At the same time, their application in ruminants' nutrition enables the sustainable use of high-added value bioactive ingredients inside food chains that improve milk and meat quality characteristics and fortify consumer health, without negative effects on rumen function, metabolism and productivity.

Abstract: The olive oil industry has a leading position in the Mediterranean countries, resulting in the production of considerable quantities of the respective by-products (OB) that constitute an important environmental issue. OB contain valuable nutrients and bioactive components that can be re-used under the bioeconomy strategy, and several chemical, physical, and biological processes have been evaluated with the intention to improve their nutritional value. One feasible application of OB is their incorporation in the diets of livestock and especially ruminants due to their high fiber content. As indicated by numerous studies, OB dietary supplementation increases the levels of monounsaturated fatty acids (MUFAs) and decreases that of saturated fatty acids (SFAs) in the milk and meat of ruminants with beneficial effects for consumers' health. At the same time, environmental impact and feeding costs are reduced without detrimental effects on ruminal fermentation, nutrients utilization, growth performance, carcass traits, milk yield and composition.

Keywords: olive by-products; ruminants; milk quality; monounsaturated fatty acids; meat quality

Citation: Tzamaloukas, O.; Neofytou, M.C.; Simitzis, P.E. Application of Olive By-Products in Livestock with Emphasis on Small Ruminants: Implications on Rumen Function, Growth Performance, Milk and Meat Quality. *Animals* **2021**, *11*, 531. https://doi.org/10.3390/ani11020531

Academic Editor: Adriana Bonanno
Received: 20 January 2021
Accepted: 11 February 2021
Published: 18 February 2021

1. Introduction

The traditional Mediterranean diet has beneficial effects on the health status of populations in South Europe. One of its main components is olive oil, which provides a high monounsaturated to saturated fat ratio in the diet [1]. Cultivation of olive (*Olea europaea*) tree is strongly related to the economy of the Mediterranean countries. In 2014, Spain, Italy, and Greece produced more than 1.74, 0.30, and 0.21 million tons of olive oil, respectively [2]. However, olive oil extraction generates a considerable quantity of by-products (pulp, olive kernels, skin, and water) that are potential environmental pollutants due to their high organic load [3]; approximately 800 g of olive cake is obtained from each kg of olives [4]. Incorporation of these by-products into the diets of livestock might mitigate the environmental burden induced by their disposal and minimize the costs related to waste management and animal feeding, since animals become less dependent on conventional feeds such as cereal grains that can be consumed by humans [5]. In addition, their application according to bioeconomy principles enables the sustainable use of high-added value ingredients—nutraceuticals—inside the food chain, protects natural resources, and mitigates climate changes [6].

The phenolic compounds of olive products have been associated with improved cardiovascular health, low cholesterol levels, and increased longevity. A variety of substances with proven antioxidant and radical scavenging activity, such as hydroxytyrosol, oleuropein, tyrosol, caffeic acid, p-cumaric acid, verbascoside, and elenolic acid, is also contained in olive cake [7]. The direct inclusion of olive by-products in pasta and baked products improves not only their antioxidant capacity, but also their sensory attributes and shelf-life, due to their content in polyunsaturated fatty acids, phenolic compounds, and dietary fiber [8,9].

Furthermore, addition of olive by-product (OB) in pig diets could serve as a feasible approach to reducing production costs, especially in the Mediterranean region, while the quality and antioxidant capacity of the derived meat is maintained [10], as indicated by the reduction in the levels of saturated fatty acids (SFAs) and the increase in that of unsaturated fatty acids (MUFAs and PUFAs) [6,11,12]. In broilers, meat MUFAs levels are increased [13] and growth parameters [14] and carcass traits [15] are improved as an effect of OB dietary inclusion. Similar findings have been shown in laying hens, since levels of cholesterol and SFAs are reduced and those of MUFAs and PUFAs are increased in egg yolk after OB dietary supplementation [16,17], without negative effects on productive performance at a level of up to 9% [18] or 16% [16].

The market continuously seeks alternative ways of improving the health benefits and technological properties of dairy and meat products derived from ruminants. The World Health Organization [19] suggested a reduction in the intake of SFAs due to their hypercholesterolemic and thrombogenic effects that are correlated with an increased risk for cardiovascular diseases. In this framework, researchers have focused their attention on reducing SFAs content and improving nutritional properties of milk and meat through the enrichment of diets with agro-industrial by-products rich in unsaturated fatty acids, such as the olive cake.

2. Olive By-Products as Non-Conventional Feed Resources

Upon oil extraction, two fractions are generated: a solid residue that is generally known as crude olive cake (OC) or olive pomace and a liquid one that refers to olive mill wastewater. OC is the mixture of the olive kernel shell (or stone or pit) crushed into fragments, the skin, and the crushed grape pulp. OC chemical composition greatly varies due to the proportion of the aforementioned solid components but also the year, the geographic origin, the olive variety, culture conditions, and possible soil contamination. It can be further categorized to crude or extracted (or exhausted or defatted or depleted) OC based on residual oil content, to fresh or dry OC based on its moisture levels and to partly destoned or crude OC based on the stone removal or not [5,6]. For instance, the crude OC obtained by mechanical extraction contains residual oil and stones. The exhausted olive cake is the residue obtained after oil extraction from the crude olive cake by a solvent, usually hexane, and the partly destoned olive cake is the result of partly separating the stone (or kernel shell) from the pulp by screening or ventilation [20]. OC could be obtained according to the applied method of oil centrifugation process (three or two phase OC). The latter one, of two phase extraction, is a more efficient and environmentally friendly procedure, since the production of olive mill wastewater is minimized. Furthermore, other by-products of olive oil extraction are the olive pulp (OP), obtained when stones are completely separated from OC during oil extraction, and the olive leaves (OL), which refers to a mixture of leaves and branches from the pruning of olive trees and the cleaning of olives prior to oil extraction [5,6].

OB are the most abundant agro-industrial by-products in the Mediterranean basin and could be applied in animal and especially ruminant diets as fresh, ensiled, dried, or a component of concentrated pellets and multi-nutrient feed blocks. Their nutritional value (fiber content, crude protein, and oil) varies according to the cultivation conditions (geographic origin, year, season, etc.), the method of oil extraction (three or two phase centrifugation), form (crude, exhausted, partly-destoned, etc.), the preservation methods (drying, ensiling, etc.), and the storage conditions and time [5,21,22] (Table 1). OC digestibility is variable, depending on its type. However, apparent digestibility of organic matter and crude protein is low (0.20–0.50) due to a low nitrogen solubility and high levels of acid detergent insoluble nitrogen (70–75% of total nitrogen), while ether extract is highly digested (0.60–0.90), regardless of type of olive cake and processing method [5].

Table 1. Composition of the different olive by-products for feeding trials.

Olive by-Product	Dry Matter (%)	Ether Extract (%)	Crude Protein (%)	NDF (%)	SFAs (%)	MUFAs (%)	PUFAs (%)	Total Phenols (%)	References
Olive cake (OC)	85–91	4–9	8–11	60–66	18–25	68–70	7–12	14	[5,23–25]
Partly destoned OC	89–92	1–5	3–13	54–68	47	45	8	0.65 mg GAE */g	[26,27]
Olive mill wastewater	-	0.2–1	-	-	-	-	-	0–1.2	[28]
Olive leaves (OL)	95	3–4	10–13	30–40	40	22	38	25	[5,23,29]

* gallic acid equivalent.

Olive by-products have traditionally been used by farmers in Mediterranean areas, but there are some limitations regarding their use as non-conventional feed resources for livestock [6]. These confinements are related to the fact that they contain low available protein content (6.6–9.9%), high ether extract (10–30%), neutral detergent fiber (23–73%) (NDF), and acid detergent lignin (12–37%) (ADL) and compounds such as phytic acid, polyphenols, and tannins that inhibit rumen cellulolytic activity and negatively affect OB palatability and digestibility [5,21,30]. Among them, the high lignin content constitutes the main obstacle towards better utilization, while the tannin content is less than 1% of DM [20]. Moreover, although they are considered as a good source of energy, this high energy content may reduce the animals' total feed intake [6]. Additionally, they contain high fat levels, making them a good supplement for a balanced diet, but fat-rich by-products should be limited to a certain percentage (10% of total diet at most, although 5% is usually recommended). As a result, participation of olive by-products in the diets of both monogastrics and ruminants is limited (5–15%) due to high content of fiber and fat, respectively [6,31]. Nasopoulou and Zabetakis [32] concluded that a moderate intake of OB does not affect growth and improves the fatty acid profile of animal products by reducing saturated and increasing unsaturated ones in both meat and milk.

Appropriate methods of collection, transportation, and processing should be implemented in order to reduce the costs and improve the nutritive value of olive cake [33]; for example, ensiling is a low cost method of preserving olive cake without the need of additives or the cost of drying/pelleting [34–36]. There are several chemical, physical, or biological procedures that can be used to increase the protein content and minimize the anti-nutritional factors (phytic acid, polyphenols, and tannins) in olive cake. In general, destoning of olive cake improves its DM, NDF, crude protein digestibility, and nutritive value [37]. Ammonia or soda treatment and addition of molasses appeared to improve palatability and nutritive value of olive cake [38,39]. At the same time, the palatability of ensiled olive cake is very high, as observed in several studies in ruminants. High intake of olive cake has been reported in ruminants [40], whereas other reports shows that heifers can consume higher amounts of ensiled olive cake compared to lams and kids (heifers 58, lambs 34, kids 28 g/kg of $W^{0.75}$) [41]. Moreover, ensiling with urea (4–5%) increased crude protein level of olive cake, but its digestibility was not improved [42,43]. On the other hand, solid-state fermentation using selected filamentous fungi is a promising biological technique that improves olive cake nutritional value and is characterized by its low cost, simplicity, and efficiency. At the same time, a more stable product is obtained, the requirements for energy are less, and smaller levels of effluents are produced compared to submerged fermentation systems [44].

3. Effects of Olive By-Products Dietary Supplementation in Ruminants

3.1. Effects of Olive By-Products Dietary Supplementation on Rumen Microbiota and Fermentation Characteristics

Olive by-products generally have low digestibility and palatability, and their high polyphenol content (particularly tannins) could decrease protein availability and microbial protein synthesis in ruminants due to the inhibitory action of polyphenols on the extracellular enzymes secreted by the ruminal microflora [45,46]. In detail, the inclusion of stoned olive cake (SOC) into the diet inhibited in vitro rumen biohydrogenation of C18 unsaturated fatty acids, resulting in a decrease in the stearic acid and an increase in vaccenic acid concentration, a fact that is possibly associated with differences in the microbial populations and activities; depression in the populations of *Butyrivibrio proteoclasticus, Neisseria weaveri* and *Ruminobacter amylophilus* [45]. Moreover, the inclusion of crude two-stage olive cake (10–12%) in feed blocks increased vaccenic acid production, since it is rich in oleic and linoleic fatty acids. The isomerization of linoleic acid as well as the desaturation of vaccenic acid in both rumen and mammary gland led to goat milk with high levels of rumenic acid and total conjugated linoleic acid (CLA) [47]. Additionally, the replacement of forage by crude olive cake at the level of 33% (or 16.6% of the total diet) in dairy sheep did not affect volatile fatty acids, ammonia production, microbial growth, bacterial diversity, protozoal, fungi, or archaea abundance, although pH and butyrate proportions were increased [48]. The inclusion of olive cake obtained with a two-stage (135 g/kg DM) or three-stage (112.5 g/kg DM) olive milling in the ewe diets indicated increased contents of α-linolenic and rumenic acids in rumen liquor (RL), respectively, while there was no diet effect on the overall composition of the RL microbiota among treatments [49]. However, in the same study, significant differences were observed for six bacterial taxa between control and treated groups. More specifically, the RL microbiota of animals fed the olive cake diets showed reduced concentration of *Anaerovibrio* genus, a result that could lead to a reduction in lipolysis, and thereby lowering the amount of PUFA that are available for biohydrogenation [49].

Incorporation of second-extraction pitted and dehydrated olive cake into the diets of Friesian steers at the level of 10–20% (on DM basis) did not influence rumen fermentation variables (pH, concentrations of ammonia and volatile fatty acids (VFA), and molar proportions of the different acids) [50]. In small ruminants, the inclusion of two-stage dried olive cake in the diet resulted in an increase of condensed tannins. Ruminal VFA concentration in goats and wethers increased, and ammonia concentration decreased. The inclusion of two-stage dried olive cake decreased urinary allantoin excretion only in wethers, indicating a greater sensitivity of wethers than of goats to olive cake condensed tannins [46] and a higher degradative efficiency of rumen microorganisms with proteolytic and cellulolytic activity in goats than in sheep [31]. Ruminal fermentation parameters (pH, VFA levels, and methane emissions) and nutrients utilization were also not disturbed in dairy goats after the partial replacement of the 20% forage by crude olive cake silages supplemented with sunflower oil at the level of 2% [51] or after the partial replacement of concentrates with a mixture of corn dried distillers' grains containing 18% solubles, 18% dry citrus pulp, and 8% exhausted olive cake (as-fed basis) [52]. The discrepancy observed in the literature is also possibly related to the different olive varieties, different oil extraction procedures used, and mainly due to the interaction of olive by-products with other dietary components.

3.2. Effects of Olive By-Products Dietary Supplementation on Milk Products

3.2.1. Dairy Cows, Buffaloes, and Camels

The partial replacement of roughage (8%) and concentrate (5%) in the total mixed ration by dried olive cake (DOC) [53] or its inclusion at the level of 5.6% (on DM basis) [54] did not affect milk yield and composition in dairy cows. Dried, partially destoned, olive cake dietary supplementation of dairy cows at the level of 15% (of DM) generally did not affect milk and cheese yield and chemical profile, but the nutritional properties of the derived cheese were also improved, since an increase in MUFAs and PUFAs and a decrease in SFAs were also observed. Moreover, atherogenic and thrombogenic indices were reduced, and oleic and CLA contents were increased as a result of the decreased biohydrogenation rate of oleic and linoleic intermediate by *Butyrivibrio* genus and *B. proteoclasticus* [55]. In buffaloes, no effect of DOC on milk yield, composition, and coagulation parameters was observed, whereas the oxidative stability and the dietetic—nutritional characteristics of the milk (increased MUFAs, PUFAs, unsaturated/saturated (UFA/SFA) ratio and decreased atherogenic and thrombogenic indices) were improved as an effect of tocopherols, retinol, and hydroxytyrosol that are presented in DOC [56].

Dietary supplementation with DOC at the level of 10% (on DM basis) did not influence milk yield and composition in dairy cows, with the exception of milk protein content that was increased [57]. An increase in MUFAs and a decrease in SFAs of milk and cheese were observed, whereas no effect of DOC on PUFAs was found. In detail, DOC dietary inclusion reduced palmitic acid and atherogenic and thrombogenic indices, while increasing oleic, vaccenic, stearic, and CLA in both milk and cheese of dairy cows [57]. Crude olive cake as a replacer of barley in the camel diets did not also affect milk yield, fat, or protein content but increased medium-chain fatty acids levels [58].

At the same time, the substitution of forages with 10% (DM) of ensiled olive cake had no effect on bovine milk yield and composition apart from an increase in milk fat yield. At the same time, a significant reduction in the content of SFAs and the atherogenic index was reported, whereas increased levels of long-chain and monounsaturated fatty acids as well as individual fatty acids like stearic, oleic, the sum of C18:1 trans-10 and trans-11 acids, and CLA (cis-9, trans-11 C18:2) were observed in milk and Halloumi cheese [36]. Moreover, Chaves et al. [59] concluded that inclusion of olive cake, conserved as silage, as a replacer of corn silage in the diet of lactating cows up to 15% (dry basis) does not alter milk production or its composition and feed efficiency.

3.2.2. Dairy Ewes and Goats

In dairy ewes, MUFAs, oleic acid, n-6/n-3 PUFAs, and UFA/SFA ratio were increased without negative effects on chemical composition and clotting properties after the incorporation of olive cake at the level of 10–25%, leading to the improvement of the dietetic-nutritional characteristics of milk and cheese and the decrease of the atherogenic and thrombogenic indices [60,61]. Other researchers that used the same levels of olive cake (20% on DM basis) did not observe a significant effect on milk yield and composition in Awassi ewes [62]. Additionally, no diet effect on milk yield and composition was observed in a recent study conducted on ewes after the inclusion of olive cake produced with a two (135 g/kg DM) or three (112.5 g/kg DM) stage milling process, while an enrichment of milk fat with α-linolenic and oleic acids was reported in both olive cake supplemented groups [49]. On the other hand, the partial replacement of conventional concentrates by a mixture containing by-products like exhausted olive cake (80 g/kg as fed-basis) significantly increased milk protein content in dairy goats and improved the quality of goat milk by significantly increasing the levels of PUFAs, CLA, and linoleic acid, decreasing at the same time those of SFAs and the n-3 to n-6 ratio [52]. The inclusion of olive cake at a higher level (30% on DM basis) did not influence milk yield and composition, while MUFAs content in milk and the derived yoghurt and cheese was improved in Awassi dairy ewes [63]. However, the same authors in a recent experiment found that the replacement of forage and concentrate at the level of 30% by olive cake reduced milk yield (−10%) and protein and increased MUFAs content in ovine milk and attributed the observed discrepancies in the increased sample size of the latter study [64].

No significant effects on milk yield and composition of dairy ewes were demonstrated after the application of olive cake as a silage (fat: 115 g/kg DM) at the level of 100 g/kg DM (as a replacement for equivalent amounts of grass hay) [65]. The strategy of partial replacement of conventional forage by ensiled olive cake (with sunflower oil) at the level of 200 g/kg DM is also a valuable nutritional strategy in dairy goats, since it improves animal energy balance and microbial protein synthesis while ruminal fermentation, nutrients utilization, milk yield, and composition are not compromised [51]. Partial replacement of conventional roughages by ensiled crude olive cake (50%) did not affect milk yield in Chios ewes and Damascus goats and significantly increased milk fat content only in dairy ewes [34]. Moreover, the substitution of forages with 72 or 142 g/kg DM of ensiled olive cake reduced SFAs and increased the unsaturated and monounsaturated lipids as well as individual fatty acids like rumenic and linoleic acids in milk fat of Chios ewes with positive effects on human health, while the cholesterol content was not affected [36]. The application of an ensiled-mixture consisting of crude olive cake, orange pulp, and wheat straw increased MUFAs levels, UFA/SFA, and n-6/n-3 PUFAs ratio in milk of Comisana ewes [66].

Alloueedat et al. [67] showed that a mix of alternative feeds including olive cake supplemented at the level of 20–40% (of DM) could be used in ewe diets to mitigate production cost without negatively affecting intake, milk yield and composition, digestibility, animal welfare, and health. Moreover, feed blocks that contained two-stage olive cake at the level of 10–12% could be used as an alternative to reduce half of the amount of concentrate without detrimental effects on nutrient utilization, nitrogen balance, and energy efficiency and milk composition in dairy goats. Although a decrease in milk yield was observed, milk quality was improved, since an increase in CLA content and a decrease in SFAs levels and atherogenicity index were shown [47]. The aforementioned studies regarding the effects of OB dietary supplementation on milk products of ruminants are summarized in Table 2.

Table 2. The effects of olive by-products dietary supplementation on milk products of ruminants.

Olive By-Product	Level	Animal	Effects	Reference
PDOC	200 g/kg	ewes	No effect on clotting properties. Increase of MUFA, PUFA, n−6/n−3 ratio, UFA/SFA of milk	[60]
PDOC	98–244 g/kg	ewes	Increase of MUFA, PUFA, n−6/n−3 ratio, UFA/SFA of milk and cheese	[61]
Dried OC	300 g/kg	ewes	No effect on milk yield and composition. Increase of MUFA in milk, cheese, and yogurt	[63]
Dried OC (as a replacer of forages and concentrates)	30%	ewes	Increase of MUFA. Decrease of milk yield and protein yield	[64]
OC	20%	ewes	No effect on milk yield and composition	[62]
OCS	100 g/kg	ewes	No effect on milk yield and composition	[65]
OCS with sunflower oil	200 g/kg	goats	No effect on milk yield, milk composition, ruminal fermentation, and nutrients utilisation. Increase of energy balance and microbial protein synthesis	[51]
OCFB	10–12%	goats	No effect on nutrient utilisation, nitrogen balance, energy efficiency, and milk composition. Decrease of milk yield. Increase of CLA content. Decrease of SFA and atherogenic index	[47]
OCS (as a replacer of forages)	50%	Ewes and goats	No effect of ewe and goat milk yield and composition apart from an increase in fat content of ewe milk	[34]

Table 2. *Cont.*

Olive By-Product	Level	Animal	Effects	Reference
Ensiled mixture containing OC	30%	ewes	Increase of MUFA, n−6/n−3 PUFA, and UFA/SFA	[66]
OC (as part of a mix of alternative feedstuffs—AF)	20–40% of AF	ewes	No effect on feed intake, milk yield and composition, digestibility, animal welfare, and health	[67]
OCS	72–142 g/kg	ewes	No effect on the cholesterol content of milk. A decrease of SFAs and an increase of UFAs, MUFAs, CLA, and linoleic acid	[35]
Two-stage and three-stage OC	112.5–135 g/kg	ewes	No effect on milk yield and composition. Increase of α-linolenic and oleic acids in milk	[49]
EOC (as part of a mix of byproducts replacing concentrates)	80 g/kg	goats	No effect on milk yield and composition except for an increase in protein content. No effect on nutrient apparent digestibility, urine N excretion, N utilization. A decrease on SFAs and n−3/n−6. An increase of PUFAs, CLA, and linoleic acid.	[52]
DSOC (as a replacer of forages and concentrates)	13%	cows	No effect on milk yield and composition	[53]
DSOC	5.6%	cows	No effect on milk yield and composition	[54]
DOC	10%	cows	No effect on milk yield and composition apart from an increase in protein content. Increase of MUFA, oleic and vaccenic acids, C-18 FA, and CLA of milk and cheese	[57]
OCS	10%	cows	No effect on milk yield and composition apart from an increase in fat yield. Increase of MUFA, oleic acid and CLA of both milk and Hamoumi cheese. Decrease of SFA and atherogenic index of milk and Hamoumi cheese	[36]
OCS	15%	cows	No effect on milk yield, milk composition, and feed efficiency	[59]
DPDOC	15%	cows	No effect on yield of milk and cheese and chemical profile of cheese. Increase of MUFA, PUFA, oleic acid, and CLA. Decrease of SFA, atherogenic, and thrombogenic indices of cheese	[55]
DSOC (as a replacer of concentrates)	15.50 %	buffaloes	No effect on milk yield, milk composition, and coagulation parameters	[56]
Crude OC (as a replacer of forages)	3 kg	camels	No effect on milk yield and composition. Increase of MUFA	[58]

CLA: conjugated linoleic acid, DOC: dried olive cake, MUFAS: monounsaturated fatty acids, OCFB: feed blocks that contained olive cake, OCS: olive cake silage, PDOC: partly destoned olive cake, DSOC: dried stoned olive cake, DOC: dried olive cake, DPDOC: dried partially destoned olive cake, EOC: exhausted olive cake; PUFAs: polyunsaturated fatty acids, SFAs: saturated fatty acids, UFAs: unsaturated fatty acids.

3.3. Effects of Olive By-Products Dietary Supplementation on Growth Performance, Carcass Traits, and Meat Quality Characteristics

3.3.1. Beef Cattle

Although the majority of the available literature concerning the effects of olive by-products refers to small ruminants, there are also some recent studies that deal with their effects in beef cattle. Inclusion of up to 20% second-extraction pitted and dehydrated olive cake (DM basis) in the diet did not affect growth performance (final body weight and average daily gain) and rumen parameters (pH, ammonia, and volatile fatty acids) of Friesian steers [50]. However, dried partially destoned olive cake (DPDOC) dietary supplementation of Limousin bulls at the level of 7.5–15% (of DM) increased final body weight, average daily gain, carcass traits, intramuscular fat content, and meat yellowness. Moreover, DPDOC inclusion at the level of 15% reduced meat cooking loss, shear force value, and SFAs levels and increased PUFAs and MUFAs content, UFA/SFA and n-6/n-3 PUFAs ratio, and oleic acid levels [68]. Finally, incorporation of calcium soap of olive oil into the diets of Blonde D'Aquitane steers at the level of 4.8% did not affect growth performance, carcass traits, and meat sensory attributes [69].

3.3.2. Sheep

An acceptable level of performance in lambs could be assured when olive cake is used as a part of the basal diet (12.2%) [70]. At the same time, the addition of higher levels of olive cake (15%) to the concentrate had no significant effect on daily gain, feed efficiency, carcass weight, and dressing percentage of lambs [71,72]. Moreover, the supplementation of lambs fed indoors or reared on a rangeland with olive cake (280 g/day) did not affect slaughter weight, carcass traits, meat yield, and quality characteristics (apart from a decrease in pH and an increase in juiciness of meat) [73]. Kotsampasi et al. [27] reached similar conclusions after the incorporation of partly destoned exhausted olive cake into lambs' diet (80–240 g/kg diet); no effect on growth performance, carcass weight, or intramuscular fatty acid profile was observed, whereas fat color, fat firmness wetness, and overall acceptability of carcasses were improved. Moreover, partly destoned exhausted olive cake could be used in lamb finishing rations at the level of 10–20% with no adverse effects on ADG and carcass characteristics, although final body weight and feed efficiency were negatively affected at the higher level (20%) [26]. Being a cheap by-product, olive cake can be utilized as an alternative feed source for lambs up to the level of 20–30% without having any detrimental effect on their growth performance and carcass traits [74–76].

Moreover, the replacement of half of dietary wheat hay with sun-dried olive cake improved weight gain and final body weight of Awassi lambs with no detrimental effects on rumen parameters, nutrients intake, or digestibility [77]. In a recent experiment of the same authors, dietary inclusion of a mix of alternative feedstuffs that contained olive cake at the level of 250 or 500 g/kg also decreased production cost with no effect on feed conversion ratio, carcass traits, or meat quality characteristics of lambs. However, at the level of 500 g/kg, nutrients intake, digestibility, and lambs' performance were negatively affected [78]. Farmers could reduce the amount of concentrate used for lambs up to 75% by using olive cake-based feed blocks (44%) as alternative supplements, since no negative effects on lamb performance are observed [79]. Taheri et al. [80] reached a similar inclusion after the inclusion of ensiled olive pulp at the level of 10–30% into the diets of ram lambs; ADG, FCR, and most carcass traits including major cuts were not adversely affected.

The replacement of the diet at the level of 44% by a mixture of agro-industrial by-products that contained exhausted olive cake did not change growth performance, pH, chemical composition, color, or texture parameters of lamb meat but increased its shelf-life (reduced lipid oxidation levels) and improved its fatty acid profile (decrease of SFAs and increase of PUFAs content) [81,82]. Furthermore, stoned olive cake dietary supplementation (35%) improved the oxidative stability of lamb meat and its combination with linseed (17% and 10%, respectively) improved the fatty acid composition of meat without compromising oxidative stability [83], with no effect on feed intake, growth performance, or carcass weight [84]. Reduced SFAs levels and increased n−6/n−3 PUFAs ratio were observed in the meat of lamb that were ad libitum fed with a silage based on olive cake and cactus pads [85]. Furthermore, polypohenols extracted from olive mill wastewaters have positive effects on kid meat, improving fatty acid profile (oleic and CLA content) and oxidative stability, suggesting the utilization of this byproduct as a source of polyphenols although this has not been investigated in any extent [86]. All the mentioned literature related to the effects of OB dietary supplementation on growth performance, carcass traits, and meat properties of lambs is summarized in Table 3.

Table 3. The effects of olive by-products dietary supplementation on growth performance, carcass traits, and meat properties of lambs.

Olive By-Product	Level	Effects	Reference
COC	280 g/kg	No effect on slaughter weight, ADG, carcass traits, meat yield, and quality parameters (CL, color), apart from a reduction in meat pH and an increase in meat juiciness.	[73]
DOC	100 g/kg	Increase of GR and FBW. No effect on carcass traits	[70]
DOC	300 g/kg	Reduction of ADG, FBW, carcass weight, and dressing percentage	[71]
DOC	150 g/kg	No effect on growth performance (FBW, ADG, and FCR),	[72]
DOC (as part of a mix of alternative feedstuffs—AF)	250 g/kg AF	No effect on growth performance (FBW, ADG, FCR), carcass traits and meat quality characteristics (pH, CL, WHC, SF, and color lightness and yellowness). A decrease of color redness was observed.	[78]
EOC	80 g/kg	No effect on meat pH, color, SF, and chemical composition. Increase of PUFAs content and n-6/n-3 ratio and reduction of SFAs and TBARS levels	[81]
FOW	50–150 g/kg	No effect on growth performance (FBW, ADG, FCR) and carcass traits	[76]
OCFB	440 g/kg	No effect on growth performance	[79]
OCS	240 g/kg	Reduction of FBW and ADG. No effect on FCR and carcass traits	[87]
OCS	300 g/kg	No effect on ADG, FCR, carcass characteristics, and major cuts, apart from a reduction in carcass dressing percentage	[80]
PDEOC	80–240 g/kg	No effect on growth performance (FBW, ADG, FCR), carcass traits, and meat chemical composition	[27]
PDEOC	100–200 g/kg	No effect on ADG and carcass major cuts. Reduction of hot carcass dressing percentage	[26]
PDOC	100–300 g/kg	No effect on growth performance (FBW, ADG, FCR) and carcass traits	[75]
SOC	350 g/kg	No effect on meat fatty acid profile. Improvement of meat oxidative stability and increase of tocopherols content	[83]
SOC	350 g/kg	No effect on growth performance (FBW, ADG), carcass traits, and meat fatty acid profile	[84]
UTOC	100–300 g/kg	No effect on FBW, ADG, and FCR	[43]

ADG: average daily gain, CL: cooking loss, COC: crude olive cake, DOC: dried olive cake, EOC: exhausted olive cake, FBW: final body weight, FCR: feed conversion rate, FOW: fermented olive wastes, GR: growth rate, OCS: olive cake silage, OCFB: feed blocks that contained olive cake, PDEOC: partly destoned exhausted olive cake, PDOC: partly destoned olive cake, PUFAs: polyunsaturated fatty acids, SF: shear force, SFAs: saturated fatty acids, SOC: stoned olive cake, TBARS: thiobarbituric acid reactive substances, UTOC: urea treated olive cake, WHC: water holding capacity.

4. Conclusions

According to the literature, it is clearly demonstrated that the incorporation of olive by-products (OB) into the diet of livestock and especially that of ruminants could serve as an advantageous strategy in the Mediterranean areas, allowing exploitation of an important costless agro-industrial by-product. Several methods of processing could be applied in order to improve OB nutritive value, with ensiling appearing as the most cost-effective. Nevertheless, the OB can safely be included up to 15–20% on DM basis without negative effects on ruminant digestion, given the variability of other diet components as indicated in several studies. Furthermore, nutritional properties of milk and meat are improved through OB dietary supplementation, without negative effects on growth performance and productivity. In detail, increased MUFAs and reduced SFAs levels are observed that are correlated with diminished hypercholesterolemic and thrombogenic effects, leading to the fortification of human health. In conclusion, OB utilization in ruminants' diets reduces production costs and mitigates environmental burden, while an improvement in the nutritional value of the derived products is observed.

Author Contributions: Conceptualization: O.T. and P.E.S.; writing—original draft preparation: M.C.N. and P.E.S.; writing—review and editing: M.C.N., O.T. and P.E.S.; supervision: P.E.S. and O.T. All authors have read and agreed to the published version of the manuscript.

Funding: This review received no external funding.

Institutional Review Board Statement: Not applicable.

Informed Consent Statement: Not applicable.

Data Availability Statement: Data sharing not applicable.

Conflicts of Interest: The authors declare no conflict of interest.

References

1. Trichopoulou, A.; Vasilopoulou, E. Mediterranean Diet and Longevity. *Br. J. Nutr.* **2000**, *84*, 205–209. [CrossRef]
2. FAOSTAT. Available online: http://faostat3.fao.org/browse/Q/QD/E (accessed on 12 December 2020).
3. Chouchene, A.; Jeguirim, M.; Khiari, B.; Zagrouba, F.; Trouvé, G. Thermal Degradation of Olive Solid Waste: Influence of Particle Size and Oxygen Concentration. *Resour. Conserv. Recycl.* **2010**, *54*, 271–277. [CrossRef]
4. Camposeo, S.; Vivaldi, G.A.; Gattullo, C.E. Ripening Indices and Harvesting Times of Different Olive Cultivars for Continuous Harvest. *Sci. Hortic.* **2013**, *151*, 1–10. [CrossRef]
5. Molina-Alcaide, E.; Yáñez-Ruiz, D. Potential Use of Olive By-Products in Ruminant Feeding: A Review. *Anim. Feed. Sci. Technol.* **2008**, *147*, 247–264. [CrossRef]
6. Berbel, J.; Posadillo, A. Review and Analysis of Alternatives for the Valorisation of Agro-Industrial Olive Oil By-Products. *Sustain.* **2018**, *10*, 237. [CrossRef]
7. Ghanbari, R.; Anwar, F.; Alkharfy, K.M.; Gilani, A.-H.; Saari, N. Valuable Nutrients and Functional Bioactives in Different Parts of Olive (*Olea europaea* L.)—A Review. *Int. J. Mol. Sci.* **2012**, *13*, 3291–3340. [CrossRef] [PubMed]
8. Araújo, M.; Pimentel, F.B.; Alves, R.C.; Oliveira, M.B.P. Phenolic Compounds from Olive Mill Wastes: Health Effects, Analytical Approach and Application as Food Antioxidants. *Trends Food Sci. Technol.* **2015**, *45*, 200–211. [CrossRef]
9. Difonzo, G.; Troilo, M.; Squeo, G.; Pasqualone, A.; Caponio, F. Functional Compounds from Olive Pomace to Obtain High-Added Value Foods—A Review. *J. Sci. Food Agric.* **2021**, *101*, 15–26. [CrossRef]
10. Tsala, A.; Mpekelis, V.; Karvelis, G.; Tsikakis, P.; Goliomytis, M.; Simitzis, P. Effects of Dried Olive Pulp Dietary Supplementation on Quality Characteristics and Antioxidant Capacity of Pig Meat. *Foods* **2020**, *9*, 81. [CrossRef]
11. Joven, M.; Pintos, E.; Latorre, M.; Suárez-Belloch, J.; Guada, J.; Fondevila, M. Effect of Replacing Barley by Increasing Levels of Olive Cake in the Diet of Finishing Pigs: Growth Performances, Digestibility, Carcass, Meat and Fat Quality. *Anim. Feed. Sci. Technol.* **2014**, *197*, 185–193. [CrossRef]
12. Ferrer, P.; Calvet, S.; García-Rebollar, P.; De Blas, C.; Jiménez-Belenguer, A.I.; Hernández, P.; Piquer, O.; Cerisuelo, A. Partially Defatted Olive Cake in Finishing Pig Diets: Implications on Performance, Faecal Microbiota, Carcass Quality, Slurry Composition and Gas Emission. *Animals* **2020**, *14*, 426–434. [CrossRef]
13. Papadomichelakis, G.; Pappas, A.; Tsiplakou, E.; Symeon, G.; Sotirakoglou, K.; Mpekelis, V.; Fegeros, K.; Zervas, G. Effects of Dietary Dried Olive Pulp Inclusion on Growth Performance and Meat Quality of Broiler Chickens. *Livest. Sci.* **2019**, *221*, 115–122. [CrossRef]
14. Herrero-Encinas, J.; Blanch, M.; Pastor, J.; Mereu, A.; Ipharraguerre, I.; Menoyo, D. Effects of a Bioactive Olive Pomace Extract from *Olea europaea* on Growth Performance, Gut Function, and Intestinal Microbiota in Broiler Chickens. *Poult. Sci.* **2020**, *99*, 2–10. [CrossRef]
15. Sayehban, P.; Seidavi, A.; Dadashbeiki, M.; Ghorbani, A.; De Araújo, W.A.G.; Durazzo, A.; Lucarini, M.; Gabrielli, P.; Omri, B.; Albino, L.F.T.; et al. Olive Pulp and Exogenous Enzymes Feed Supplementation Effect on the Carcass and Offal in Broilers: A Preliminary Study. *Agric.* **2020**, *10*, 359. [CrossRef]
16. Afsari, M.; Mohebbifar, A.; Torki, M. Effects of Dietary Inclusion of Olive Pulp Supplemented with Probiotics on Productive Performance, Egg Quality and Blood Parameters of Laying Hens. *Annu. Res. Rev. Biol.* **2014**, *4*, 198–211. [CrossRef]
17. Abd El-Samee, L.D.; Hashish, S.M. Olive Cake in Laying Hen Diets for Modification of Yolk Lipids. *J. Agric. Sci. Technol. A* **2011**, *1*, 415–421.
18. Zarei, M.; Ehsani, M.; Torki, M. Productive Performance of Laying Hens Fed Wheat-Based Diets Included Olive Pulp with or without a Commercial Enzyme Product. *Afr. J. Biotechnol.* **2011**, *10*, 4303–4312.
19. World Health Organization. *Diet, Nutrition and the Prevention of Chronic Diseases. Report of a Joint WHO/FAO Expert Consultation*; WHO Technical Report Series; World Health Organization: Geneva, Switzerland, 2003; p. 916.
20. Sansoucy, R.; Alibes, X.; Berge, P.; Martilotti, F.; Nefzaoui, A.; Zoiopoulos, P. Olive By-Products for Animal Feed. FAO Animal Production and Health. *Food Agric. Organ. United Nations* **1985**, *43*, 14–18.
21. Marcos, C.N.; De Evan, T.; García-Rebollar, P.; De Blas, C.; Carro, M.D. Influence of Storage Time and Processing on Chemical Composition and in Vitro Ruminal Fermentation of Olive Cake. *J. Anim. Physiol. Anim. Nutr.* **2019**, *103*, 1303–1312. [CrossRef]

22. Yansari, A.T.; Sadeghi, H.; Ansari-Pirsarai, Z.; Mohammad-Zadeh, H. Ruminal Dry Matter and Nutrient Degradability of Dif-Ferent Olive Cake By-Products after Incubation in the Rumen Using Nylon Bag Technique. *Int. J. Agric. Biol.* **2007**, *9*, 439–442.
23. Abbeddou, S.; Rischkowsky, B.; Richter, E.; Hess, H.; Kreuzer, M. Modification of Milk Fatty Acid Composition by Feeding Forages and Agro-Industrial Byproducts from Dry Areas to Awassi Sheep. *J. Dairy Sci.* **2011**, *94*, 4657–4668. [CrossRef] [PubMed]
24. El-Moneim, A.E.A.; Sabic, E.M. Beneficial Effect of Feeding Olive Pulp and Aspergillus Awamori on Productive Performance, Egg Quality, Serum/Yolk Cholesterol and Oxidative Status in Laying Japanese Quails. *J. Anim. Feed. Sci.* **2019**, *28*, 52–61. [CrossRef]
25. Tortuero, F.; Rioperez, J.; Rodríguez, M. Nutritional Value for Rabbits of Olive Pulp and the Effects on Their Visceral Organs. *Anim. Feed. Sci. Technol.* **1989**, *25*, 79–87. [CrossRef]
26. Tufarelli, V.; Introna, M.; Cazzato, E.; Mazzei, D.; Laudadio, V. Suitability of Partly Destoned Exhausted Olive Cake as by-Product Feed Ingredient for Lamb Production. *J. Anim. Sci.* **2013**, *91*, 872–877. [CrossRef] [PubMed]
27. Kotsampasi, B.; Bampidis, V.; Tsiaousi, A.; Christodoulou, C.; Petrotos, K.; Amvrosiadis, I.; Fragioudakis, N. Effects of Dietary Partly Destoned Exhausted Olive Cake Supplementation on Performance, Carcass Characteristics and Meat Quality of Growing Lambs. *Small Rumin. Res.* **2017**, *156*, 33–41. [CrossRef]
28. Zbakh, H.; El Abbassi, A. Potential Use of Olive Mill Wastewater in the Preparation of Functional Beverages: A Review. *J. Funct. Foods* **2012**, *4*, 53–65. [CrossRef]
29. Hukerdi, Y.J.; Nasri, M.F.; Rashidi, L.; Ganjkhanlou, M.; Emami, A. Effects of Dietary Olive Leaves on Performance, Carcass Traits, Meat Stability and Antioxidant Status of Fattening Mahabadi Male Kids. *Meat Sci.* **2019**, *153*, 2–8. [CrossRef]
30. Alcaide, E.M.; Ruiz, D.Y.; Moumen, A.; García, I.M. Chemical Composition and Nitrogen Availability for Goats and Sheep of Some Olive By-Products. *Small Rumin. Res.* **2003**, *49*, 329–336. [CrossRef]
31. García, A.M.; Moumen, A.; Ruiz, D.Y.; Alcaide, E.M. Chemical Composition and Nutrients Availability for Goats and Sheep of Two-Stage Olive Cake and Olive Leaves. *Anim. Feed. Sci. Technol.* **2003**, *107*, 61–74. [CrossRef]
32. Nasopoulou, C.; Zabetakis, I. Agricultural and Aquacultural Potential of Olive Pomace A Review. *J. Agric. Sci.* **2013**, *5*, 116–127. [CrossRef]
33. Ajila, C.M.; Brar, S.K.; Verma, M.; Tyagi, R.D.; Godbout, S.; Valéro, J.R. Bio-Processing of Agro-Byproducts to Animal Feed. *Crit. Rev. Biotechnol.* **2011**, *32*, 382–400. [CrossRef]
34. Hadjipanayiotou, M. Feeding Ensiled Crude Olive Cake to Lactating Chios Ewes, Damascus Goats and Friesian Cows. *Livest. Prod. Sci.* **1999**, *59*, 61–66. [CrossRef]
35. Symeou, S.; Tsiafoulis, C.G.; Gerothanassis, I.P.; Miltiadou, D.; Tzamaloukas, O. Nuclear Magnetic Resonance Screening of Changes in Fatty Acid and Cholesterol Content of Ovine Milk Induced by Ensiled Olive Cake Inclusion in Chios Sheep Diets. *Small Rumin. Res.* **2019**, *177*, 111–116. [CrossRef]
36. Neofytou, M.; Miltiadou, D.; Sfakianaki, E.; Constantinou, C.; Symeou, S.; Sparaggis, D.; Hager-Theodorides, A.; Tzamaloukas, O. The Use of Ensiled Olive Cake in the Diets of Friesian Cows Increases Beneficial Fatty Acids in Milk and Halloumi Cheese and Alters the Expression of SREBF1 in Adipose Tissue. *J. Dairy Sci.* **2020**, *103*, 8998–9011. [CrossRef]
37. Sadeghi, H.; Yansari, A.T.; Ansari-Pirsarai, Z. Effects of Different Olive Cake by Products on Dry Matter Intake, Nutrient Digestibility and Performance of Zel Sheep. *Int. J. Agric. Biol.* **2009**, *11*, 39–43.
38. Amici, A.; Verna, M.; Martillotti, F. Olive By-Products in Animal Feeding: Improvement and Utilization. *Options Méditerranéennes* **1991**, *16*, 149–152.
39. Weinberg, Z.; Chen, Y.; Weinberg, P. Ensiling Olive Cake with and without Molasses for Ruminant Feeding. *Bioresour. Technol.* **2008**, *99*, 1526–1529. [CrossRef] [PubMed]
40. Nefzaoui, A.; Vanbelle, M. Effects of Feeding Alkali-Treated Olive Cake on Intake, Digestibility and Rumen Liquor Parameters. *Anim. Feed. Sci. Technol.* **1986**, *14*, 139–149. [CrossRef]
41. Hadjipanayiotou, M.; Koumas, A. *Performance of Sheep and Goats on Olive Cake Silages. Technical Bulletin*; Agricultural Research Institute: Nicosia, Cyprus, 1996; p. 10.
42. Aboul-Fotouh, G.E.; Kamel, M.; Rady, H.; Mahfouz, H. Effect of Olive Cake Level in Sheep Ration without or with Urea or Yeast on Digestibility Coefficients and Nutritive Value. *Egypt. J. Nutr. Feeds* **2013**, *16*, 225–233.
43. Jassim, A.R.; Awadeh, F.; Abodabos, A. Supplementary Feeding Value of Urea-Treated Olive Cake When Fed to Growing Awassi Lambs. *Anim. Feed. Sci. Technol.* **1997**, *64*, 287–292. [CrossRef]
44. Chebaibi, S.; Grandchamp, M.L.; Burgé, G.; Clément, T.; Allais, F.; Laziri, F. Improvement of Protein Content and Decrease of Anti-nutritional Factors in Olive Cake by Solid-State Fermentation: A Way to Valorize This Industrial By-Product in Animal Feed. *J. Biosci. Bioeng.* **2019**, *128*, 384–390. [CrossRef]
45. Pallara, G.; Buccioni, A.; Pastorelli, R.; Minieri, S.; Mele, M.; Rapaccini, S.; Messini, A.; Pauselli, M.; Servili, M.; Giovannetti, L.; et al. Effect of Stoned Olive Pomace on Rumen Microbial Communities and Polyunsaturated Fatty Acid Biohydrogenation: An in Vitrostudy. *BMC Veter. Res.* **2014**, *10*, 1–15. [CrossRef]
46. Ruiz, D.R.Y.; Moumen, A.; García, A.I.M.; Alcaide, E.M. Ruminal Fermentation and Degradation Patterns, Protozoa Population, and Urinary Purine Derivatives Excretion in Goats and Wethers Fed Diets Based on Two-Stage Olive Cake: Effect of PEG supply1. *J. Anim. Sci.* **2004**, *82*, 2023–2032. [CrossRef]
47. Molina-Alcaide, E.; Morales-García, E.; Martin-Garcia, A.I.; Ben Salem, H.; Nefzaoui, A.; Sanz-Sampelayo, M.; Morales-García, Y.E. Effects of Partial Replacement of Concentrate with Feed Blocks on Nutrient Utilization, Microbial N Flow, and Milk Yield and Composition in Goats. *J. Dairy Sci.* **2010**, *93*, 2076–2087. [CrossRef] [PubMed]

48. García-Rodríguez, J.; Mateos, I.; Saro, C.; González, J.S.; Carro, M.D.; Ranilla, M.J. Replacing Forage by Crude Olive Cake in a Dairy Sheep Diet: Effects on Ruminal Fermentation and Microbial Populations in Rusitec Fermenters. *Animals* **2020**, *10*, 2235. [CrossRef] [PubMed]
49. Mannelli, F.; Cappucci, A.; Pini, F.; Pastorelli, R.; Decorosi, F.; Giovannetti, L.; Mele, M.; Minieri, S.; Conte, G.; Pauselli, M.; et al. Effect of Different Types of Olive Oil Pomace Dietary Supplementation on the Rumen Microbial Community Profile in Comisana Ewes. *Sci. Rep.* **2018**, *8*, 1–11. [CrossRef]
50. Estaún, J.; Dosil, J.; Al Alami, A.; Gimeno, A.; De Vega, A. Effects of Including Olive Cake in the Diet on Performance and Rumen Function of Beef Cattle. *Anim. Prod. Sci.* **2014**, *54*, 1817–1821. [CrossRef]
51. Arco-Pérez, A.; Ramos-Morales, E.; Yáñez-Ruiz, D.; Abecia, L.; Martín-García, A. Nutritive Evaluation and Milk Quality of Including of Tomato or Olive By-Products Silages with Sunflower Oil in the Diet of Dairy Goats. *Anim. Feed. Sci. Technol.* **2017**, *232*, 57–70. [CrossRef]
52. Marcos, C.; Carro, M.; Yepes, J.F.; Haro, A.; Romero-Huelva, M.; Molina-Alcaide, E. Effects of Agroindustrial by-Product Supplementation on Dairy Goat Milk Characteristics, Nutrient Utilization, Ruminal Fermentation, and Methane Production. *J. Dairy Sci.* **2020**, *103*, 1472–1483. [CrossRef] [PubMed]
53. Cibik, M.; Keles, G. Effect of Stoned Olive Cake on Milk Yield and Composition of Dairy Cows. *Revue Méd. Vét.* **2016**, *167*, 154–158.
54. Zilio, D.M.; Bartocci, S.; Di Giovanni, S.; Servili, M.; Chiariotti, A.; Terramoccia, S. Evaluation of Dried Stoned Olive Pomace as Supplementation for Lactating Holstein Cattle: Effect on Milk Production and Quality. *Anim. Prod. Sci.* **2015**, *55*, 185–188. [CrossRef]
55. Chiofalo, B.; Di Rosa, A.R.; Presti, V.L.; Chiofalo, V.; Liotta, L. Effect of Supplementation of Herd Diet with Olive Cake on the Composition Profile of Milk and on the Composition, Quality and Sensory Profile of Cheeses Made Therefrom. *Animals* **2020**, *10*, 977. [CrossRef]
56. Terramoccia, S.; Bartocci, S.; Taticchi, A.; Di Giovanni, S.; Pauselli, M.; Mourvaki, E.; Urbani, S.; Servili, M. Use of Dried Stoned Olive Pomace in the Feeding of Lactating Buffaloes: Effect on the Quantity and Quality of the Milk Produced. *Asian Australas. J. Anim. Sci.* **2013**, *26*, 971–980. [CrossRef]
57. Castellani, F.; Vitali, A.; Bernardi, N.; Marone, E.; Palazzo, F.; Grotta, L.; Martino, G. Dietary Supplementation with Dried Olive Pomace in Dairy Cows Modifies the Composition of Fatty Acids and the Aromatic Profile in Milk and Related Cheese. *J. Dairy Sci.* **2017**, *100*, 8658–8669. [CrossRef] [PubMed]
58. Faye, B.; Konuspayeva, G.; Narmuratova, M.; Serikbaeva, A.; Musaad, A.M.; Mehri, H. Effect of Crude Olive Cake Supplementation on Camel Milk Production and Fatty Acid Composition. *Dairy Sci. Technol.* **2013**, *93*, 225–239. [CrossRef]
59. Chaves, B.W.; Valles, G.A.F.; Scheibler, R.B.; Junior, J.S.; Nornberg, J.L. Milk Yield of Cows Submitted to Different Levels of Olive Pomace in the Diet. *Acta Sci. Anim. Sci.* **2020**, *43*, e51158. [CrossRef]
60. Chiofalo, B.; Liotta, L.; Zumbo, A. Administration of Olive Cake for Ewe Feeding: Effect on Milk Yield and Composition. *Small Rumin. Res.* **2004**, *55*, 169–176. [CrossRef]
61. Vargasbelloperez, E.; Vera, R.; Aguilar, C.; Lira, R.; Pena, I.; Fernández, J. Feeding Olive Cake to Ewes Improves Fatty Acid Profile of Milk and Cheese. *Anim. Feed. Sci. Technol.* **2013**, *184*, 94–99. [CrossRef]
62. Shdaifat, M.; Al-Barakah, F.; Kanan, A.; Obeidat, B. The Effect of Feeding Agricultural by-Products on Performance of Lactating Awassi Ewes. *Small Rumin. Res.* **2013**, *113*, 11–14. [CrossRef]
63. Abbeddou, S.; Rischkowsky, B.; Hilali, M.E.-D.; Hess, H.D.; Kreuzer, M. Influence of Feeding Mediterranean Food Industry by-Products and Forages to Awassi Sheep on Physicochemical Properties of Milk, Yoghurt and Cheese. *J. Dairy Res.* **2011**, *78*, 426–435. [CrossRef]
64. Abbeddou, S.; Rischkowsky, B.; Hilali, M.E.-D.; Haylani, M.; Hess, H.D.; Kreuzer, M. Supplementing Diets of Awassi Ewes with Olive Cake and Tomato Pomace: On-Farm Recovery of Effects on Yield, Composition and Fatty Acid Profile of the Milk. *Trop. Anim. Heal. Prod.* **2014**, *47*, 145–152. [CrossRef] [PubMed]
65. Cabiddu, A.; Canu, M.; Decandia, M.; Molle, G.; Pompel, R. The Intake and Performance of Dairy Ewes Fed with Different Levels of Olive Cake Silage in Late Pregnancy and Suckling Periods. In *Nutrition and Feeding Strategies of Sheep and Goats under Harsh Climates*; Ben Salem, H., Nefzaoui, A., Morand-Fehr, P., Eds.; CIHEAM-IAMZ, Options Mediterraneennes: Zaragoza, Spain, 2004; pp. 197–201.
66. Caparra, P.; Foti, F.; Scerra, M.; Postorino, S.; Vottari, G.; Cilione, C.; Scerra, V.; Sinatra, M.C. Effects of Olive Cake, Citrus Pulp and Wheat Straw Silage on Milk Fatty Acid Composition of Comisana Ewes. In *Advanced Nutrition and Feeding Strategies to Improve Sheep and Goat*; Biondi, L., Ben Salem, H., Morand-Fehr, P., Eds.; CIHEAM: Zaragoza, Spain, 2007; pp. 101–105.
67. Aloueedat, M.K.; Obeidat, B.S.; Awawdeh, M.S. Effects of Partial Replacement of Conventional with Alternative Feeds on Nutrient Intake, Digestibility, Milk Yield and Composition of Awassi Ewes and Lambs. *Animals* **2019**, *9*, 684. [CrossRef]
68. Chiofalo, V.; Liotta, L.; Presti, V.L.; Gresta, F.; Di Rosa, A.R.; Chiofalo, B. Effect of Dietary Olive Cake Supplementation on Performance, Carcass Characteristics, and Meat Quality of Beef Cattle. *Animals* **2020**, *10*, 1176. [CrossRef]
69. Castro, T.; Cabezas, A.; De La Fuente, J.; Isabel, B.; Manso, T.; Jimeno, V. Animal Performance and Meat Characteristics in Steers Reared in Intensive Conditions Fed with Different Vegetable Oils. *Animals* **2016**, *10*, 520–530. [CrossRef] [PubMed]
70. Awawdeh, M.; Obeidat, B. Effect of Supplemental Exogenous Enzymes on Performance of Finishing Awassi Lambs Fed Olive Cake-Containing Diets. *Livest. Sci.* **2011**, *138*, 20–24. [CrossRef]

71. Mioč, B.; Pavić, V.; Vnučec, I.; Prpić, Z.; Kostelic, A.; Šušić, V. Effect of Olive Cake on Daily Gain, Carcass Characteristics and Chemical Composition of Lamb Meat. *Czech, J. Anim. Sci.* **2008**, *52*, 31–36. [CrossRef]
72. Obeidat, B.S. The Effects of Feeding Olive Cake and Saccharomyces Cerevisiae Supplementation on Performance, Nutrient Digestibility and Blood Metabolites of Awassi Lambs. *Anim. Feed. Sci. Technol.* **2017**, *231*, 131–137. [CrossRef]
73. Hamdi, H.; Majdoub-Mathlouthi, L.; Picard, B.; Listrat, A.; Durand, D.; Znaïdi, I.; Kraiem, K. Carcass Traits, Contractile Muscle Properties and Meat Quality of Grazing and Feedlot Barbarine Lamb Receiving or Not Olive Cake. *Small Rumin. Res.* **2016**, *145*, 85–93. [CrossRef]
74. Awawdeh, M.S. Alternative Feedstuffs and Their Effects on Performance of Awassi Sheep: A Review. *Trop. Anim. Heal. Prod.* **2011**, *43*, 1297–1309. [CrossRef]
75. Benbati, M.; Belafqih, B.; El Otmani, S.; Mounsif, M.; Keli, A. Effect of the Level of Incorporation of Olive Cake in the Diet on Lamb Fattening Performance and Carcass Characteristics. In *Forage Resources and Ecosystem Services Provided by Mountain and Mediter-Ranean Grasslands and Rangelands*; Baumont, R., Carrère, P., Jouven, M., Lombardi, G., López-Francos, A., Martin, B., Peeters, A., Porqueddu, C., Eds.; Options Mediterraneennes: Zaragoza, Spain, 2014; pp. 261–264.
76. Christodoulou, V.; Bampidis, V.; Israilides, C.; Robinson, P.; Giouzelyiannis, A.; Vlyssides, A. Nutritional Value of Fermented Olive Wastes in Growing Lamb Rations. *Anim. Feed. Sci. Technol.* **2008**, *141*, 375–383. [CrossRef]
77. Awawdeh, M.S.; Obeidat, B.S. Treated Olive Cake as a Non-forage Fiber Source for Growing Awassi Lambs: Effects on Nutrient Intake, Rumen and Urine pH, Performance, and Carcass Yield. *Asian Australas. J. Anim. Sci.* **2013**, *26*, 661–667. [CrossRef]
78. Awawdeh, M.S.; Dager, H.K.; Obeidat, B.S. Effects of Alternative Feedstuffs on Growth Performance, Carcass Characteristics, and Meat Quality of Growing Awassi Lambs. *Ital. J. Anim. Sci.* **2019**, *18*, 777–785. [CrossRef]
79. Ben Salem, H.; Znaidi, I.-A. Partial Replacement of Concentrate with Tomato Pulp and Olive Cake-Based Feed Blocks as Supplements for Lambs Fed Wheat Straw. *Anim. Feed. Sci. Technol.* **2008**, *147*, 206–222. [CrossRef]
80. Taheri, M.R.; Zamiri, M.J.; Rowghani, E.; Akhlaghi, A. Effect of Feeding Olive-Pulp Ensiled with Additives on Feedlot Performance and Carcass Attributes of Fat-Tailed Lambs. *Trop. Anim. Heal. Prod.* **2012**, *45*, 345–350. [CrossRef]
81. De Evan, T.; Cabezas, A.; De La Fuente, J.; Carro, M.D. Feeding Agroindustrial Byproducts to Light Lambs: Influence on Growth Performance, Diet Digestibility, Nitrogen Balance, Ruminal Fermentation, and Plasma Metabolites. *Anim.* **2020**, *10*, 600. [CrossRef]
82. Evan, D.T.; Cabezas, A.; Vázquez, J.D.L.F.; Carro, M.D. Feeding Agro-Industrial By-Products to Light Lambs: Influence on Meat Characteristics, Lipid Oxidation, and Fatty Acid Profile. *Animals* **2020**, *10*, 1572. [CrossRef] [PubMed]
83. Luciano, G.; Pauselli, M.; Servili, M.; Mourvaki, E.; Serra, A.; Monahan, F.; Lanza, M.; Priolo, A.; Zinnai, A.; Mele, M. Dietary Olive Cake Reduces the Oxidation of Lipids, Including Cholesterol, in Lamb Meat Enriched in Polyunsaturated Fatty Acids. *Meat Sci.* **2013**, *93*, 703–714. [CrossRef]
84. Mele, M.; Serra, A.; Pauselli, M.; Luciano, G.; Lanza, M.; Pennisi, P.; Conte, G.; Taticchi, A.; Esposto, S.; Morbidini, L. The Use of Stoned Olive Cake and Rolled Linseed in the Diet of Intensively Reared Lambs: Effect on the Intramuscular Fatty-Acid Composition. *Animals* **2014**, *8*, 152–162. [CrossRef] [PubMed]
85. Vasta, V.; Abidi, S.; Ben Salem, H.; Nezfauoi, A.; Priolo, A. Effects of the Supplementation of Olive Cake and Cactus Pad Silage on Sheep Intramuscular Fatty Acid Composition. In *Mediterranean Livestock Production: Uncertainties and Opportunities*; Olaizola, A., Boutonnet, J.-P., Bernués, A., Eds.; Options Mediterraneanens: Zaragoza, Spain, 2008; pp. 341–344.
86. Cimmino, R.; Barone, C.M.A.; Claps, S.; Varricchio, E.; Rufrano, D.; Caroprese, M.; Albenzio, M.; De Palo, P.; Campanile, G.; Neglia, G. Effects of Dietary Supplementation with Polyphenols on Meat Quality in Saanen Goat Kids. *BMC Veter. Res.* **2018**, *14*, 181. [CrossRef]
87. Caparra, P.; Foti, F.; Cilione, C.; Scerra, M.; Vottari, G.; Chies, L. Olive Cake, Citrus Pulp and Wheat Straw Silage as an Ingredient in Lamb Diets: 1. Effects on Growth and Carcass Characteristics. *Ital. J. Anim. Sci.* **2003**, *2*, S488–S490.

animals

Article

Effects of Cornus and Its Mixture with Oregano and Thyme Essential Oils on Dairy Sheep Performance and Milk, Yoghurt and Cheese Quality under Heat Stress

Konstantinos Kalaitsidis [1], Erasmia Sidiropoulou [1], Olga Tsiftsoglou [2], Ioannis Mourtzinos [3], Thomas Moschakis [3], Zoitsa Basdagianni [4], Stylianos Vasilopoulos [1], Styliani Chatzigavriel [5], Diamanto Lazari [2] and Ilias Giannenas [1,*]

1 Laboratory of Nutrition, School of Veterinary Medicine, Aristotle University of Thessaloniki, 54124 Thessaloniki, Greece; kalaitsidisf92@gmail.com (K.K.); esidiropoulou@vet.auth.gr (E.S.); svasilopoulos@vet.auth.gr (S.V.)
2 Laboratory of Pharmacognosy, School of Pharmacy, Aristotle University of Thessaloniki, 54124 Thessaloniki, Greece; olga@tsiftsoglou.gr (O.T.); dlazari@pharm.auth.gr (D.L.)
3 Department of Food Science and Technology, School of Agriculture, Aristotle University of Thessaloniki, 54124 Thessaloniki, Greece; mourtzinos@agro.auth.gr (I.M.); tmoschak@agro.auth.gr (T.M.)
4 Department of Animal Production, School of Agriculture, Aristotle University of Thessaloniki, 54124 Thessaloniki, Greece; basdagianni@agro.auth.gr
5 ORIZON Diatrofiki, 62100 Serres, Greece; stella@orizon-ate.gr
* Correspondence: igiannenas@vet.auth.gr

Citation: Kalaitsidis, K.; Sidiropoulou, E.; Tsiftsoglou, O.; Mourtzinos, I.; Moschakis, T.; Basdagianni, Z.; Vasilopoulos, S.; Chatzigavriel, S.; Lazari, D.; Giannenas, I. Effects of Cornus and Its Mixture with Oregano and Thyme Essential Oils on Dairy Sheep Performance and Milk, Yoghurt and Cheese Quality under Heat Stress. *Animals* **2021**, *11*, 1063. https://doi.org/10.3390/ani11041063

Academic Editor: George F. W. Haenlein

Received: 24 February 2021
Accepted: 6 April 2021
Published: 8 April 2021

Simple Summary: Various plant extracts have been used as feed additives to benefit ruminant performance. In this study we investigated the effects of dietary cornus plant extract supplementation, with or without the addition of oregano and thyme essential oil, in dairy ewes. Ewes participating in this experiment were reared under thermal stress and their performance, as well as the composition of their final products (milk, feta cheese and yoghurt), were assessed. The outcome of this experiment showed that cornus plant extract, alone or in combination with oregano and thyme, favored the production of ewe's milk, along with the composition of milk and milk products.

Abstract: The effect of a diet supplemented with a novel cornus extract, enriched with essential oils of oregano and thyme, on the performance of Chios cross-bred dairy sheep was investigated during the summer period. The plant extracts were prepared using a "green" method based on aqueous extraction. A total of 45 lactating ewes were allocated into three equal groups in a randomized block design. The three groups were fed the same feed allowance, roughage based on Lucerne hay and wheat straw and a concentrate based on cereals and oil cakes (the control diet). The diet of two groups was fortified with cornus extract, with or without oregano and thyme essential oils, at a level 0.515 g of plant extract/essential oils per kg of concentrate. Individual milk yield was recorded weekly and feed refusals were recorded on a pen basis daily, during a six-week period of lactation. Milk samples were analyzed for the chemical composition of protein, fat, lactose and solids-not-fat constituents, somatic cell counts and total viable bacteria counts. Moreover, the milk of each group was used for yoghurt and Feta cheese production. The lipid oxidative stability, protein carbonyl content and fatty acid composition of milk, yoghurt and cheese samples were also evaluated. The results showed that the incorporation of novel plant extracts and essential oils increased the milk production per ewe. Dietary supplementation with cornus extracts and essential oils lowered lipid and protein oxidation in milk, yoghurt and cheese samples, compared to the control. However, diet supplementation with herbal extracts did not affect the fatty acid profile in milk, cheese and yoghurt or the serum biochemical parameters. In conclusion, dietary supplementation with cornus in combination with oregano and thyme has the potential to improve feed utilization and the performance of high-yield dairy Chios cross-bred ewes reared under heat stress.

Keywords: cornus; oregano; thyme; sheep; milk; heat stress

1. Introduction

The production of sheep milk is of great importance to the agricultural sector in Mediterranean countries, and the largest part of its volume is designated for yoghurt and cheese production [1]. Several attempts have been made to manipulate rumen efficiency in order to maximize the productivity of ruminants with respect to milk production by including various feed additives in the animals' diets, such as antibiotics, hormones, chemical growth promoters [2], enzymes and minerals [3] and plant extracts [4,5]. Feed additives such as antibiotics, hormones and other pharmaceutical substances, despite their financial benefits, have been prohibited in most countries of the world, mainly due to their potentially hazardous residues and due to the development of multi-drug resistance in bacteria [6]. The antimicrobial action of essential oils (EOs) has been known for decades and a large number of scientific studies have been carried out [7]. Recent scientific interest has focused on the addition of essential oils to ruminant nutrition in order to improve nutrient utilization efficiency and dairy product quality. Moreover, essential oils at high concentrations inhibit amino acid deamination and decrease methane production [8,9]. In view of this, the use of plant extracts and essential oils in sheep feeding should be further exploited. Additionally, milk and dairy products naturally enriched with antioxidant compounds may provide extra health benefits to the consumer [10]. Indeed, medicinal aromatic plants are used in animal nutrition and their exploitation has been greatly increased during the last decade [11].

Sheep milk is highly nutritious [1] and cheese such as feta, which is the most well-known Greek protected designation of origin product, have gained significant acceptance worldwide. It is a high-quality white-brined cheese produced from sheep milk or a mixture of sheep and goat milk ($\leq$30%) [12]. It has a salty and slightly pungent flavor [13]. Nowadays in Greece, the majority of Feta cheese is manufactured in well-organized cheese dairies using pasteurized milk and lactic acid bacteria (LAB) as starter cultures [14]. Yoghurt is another very popular fermented dairy product which has gained significant acceptance worldwide, and its nutritional and health benefits have been well known for centuries [15]. It is an excellent source of high biological value protein, essential amino acids, calcium, phosphorus, vitamins and trace minerals such as magnesium and zinc [15].

The most commonly used plants for essential oil production in animal nutrition are *Oreganum vulgare* and *Thymus vulgaris*, two plants which are widely distributed in Greece and other Mediterranean countries. Oregano and thyme essential oils have been used in animal nutrition due to their strong antimicrobial and antiparasitic action, along with their influence on polyunsaturated fatty acid protection and the antioxidant capacity delivered to the final dairy products [16]. The use of herbal thyme as a galactagogue is known to have a beneficial effect, leading to an increase in milk production [17]. Another plant commonly found in the Mediterranean area is *Cornus* spp., with a high antioxidant activity due to its large total phenolic concentration of about 220 mg/dL gallic acid equivalents [18]. The most important phenolic compounds of cornus are anthocyanins, gallic acid and ellagic acid, which all have antioxidant and antimicrobial activity [19].

The isolation of bioactive compounds from aromatic and medicinal plants is usually performed by extraction using several solvents, most of them organic, which are harmful to the environment. Thus, it is important to explore the use of extraction methods that combine eco-friendly solvents and a high yield of bioactive compounds. Green extraction methods are using aqueous solutions, thereby avoiding organic solvents such as ethanol or methanol or energy-consuming methods such as microwave extraction or high-pressure extraction [20]. The effectiveness of aqueous solutions of cyclodextrin (CD) as an eco-friendly solvent, appropriate for extracting phenolic compounds and other phytonutrients from plant materials, such as pomegranate, olive leaves, aromatic and medicinal plants or by-products of agricultural production, has been reported [21]. The addition of CD to water augments the extraction of phenolic compounds and decrease the extraction time and temperature. The extraction of essential oils from aromatic plants can be achieved by means of steam distillation, providing products with a high phenolic content, although at a relatively high cost.

A recent major issue during the last two decades concerning dairy sheep farming in the Mediterranean is the high ambient temperatures during the summer season and the first period in autumn. High ambient temperatures, combined with high direct and indirect solar radiation, low wind speed and increased relative humidity, are commonly faced by intensively reared animals that have to survive out of their thermoneutral zone (5 °C to 25 °C) [22] and maintain lactation, while possibly suffering from heat stress [23,24]. Heat stress is a major limiting factor of dairy production in hot climates [25] that is often associated with oxidative stress, defined as the presence of a large amount of reactive oxygen species, exceeding the available antioxidant capacity of animal cells [26]. Plant extracts, rich in polyphenols, may support sheep that suffer from lipoperoxidation [27].

The objective of this study was to investigate the effect of diet supplementation with a cornus plant extract alone or in combination with oregano and thyme essential oil on the milk production and composition of dairy sheep reared under heat stress conditions, the effect on animal health, as well as the effect on the nutritional value and the characteristics of two traditional dairy products—yoghurt and feta cheese. The hypothesis of this study was that the addition of bioactive plant extracts to the sheep nutrition will lead to improved milk yield, along with increased antioxidant activity in the milk, yoghurt and cheese and the capability of animals to cope better with the adverse effects of heat stress. It must be noted that there is a lack of information on the effects of cornus, oregano and thyme in dairy sheep reared during the hot seasons.

2. Materials and Methods

2.1. Ethics Guidelines of the Animal Research

This trial was carried out according to the regulations of the Greek Public Veterinary Service and approved by the Research Committee of Aristotle University of Thessaloniki, under the project numbered 99864. The health of the animals was monitored by a veterinary surgeon; animals were routinely vaccinated against enterotoxaemia clostridial infections 2 weeks before lambing, contagious agalactia, subclinical mastitis, pasteurelosis and *Brucella melitensis*. All animals had been drenched with an antiparasitic drug containing Fenbendazole 5%, and no ewe received any antimicrobial agents during lactation. All procedures were performed by the same team members on all herds throughout the project.

2.2. Experimental Design and Feeding

A total of 45 lactating ewes of the Chios cross-breed were randomly selected for twin birth from a flock consisting of 150 ewes, kept on a dairy breeding sheep farm located in the area of Polykastro (Latitude: 41°00′ N; Longitude: 22°33′ E) in the northwestern part of Kilkis in Greece. The daily mean temperature during July and August (2020) ranged between 27–34 °C. Figures 1 and 2 provide temperature data inside the building. The relative humidity was 65% and the average wind speed was one meter per second. The ewes were allocated into three equally sized treatment groups of 15 ewes, each consisting of 5 subgroups housed in a separate pen. Each group and subgroup was balanced for body weight (BW) three weeks after lambing, parity and body condition. The three treatment groups were fed isonitrogenous and isoenergetic diets. Group 1 (Control) was fed the basal ration with Lucerne hay and wheat straw and a concentrate containing maize, soybean meal and sunflower meal as the main constituents (Table 1). The diet of group 2 was supplemented with cornus extract at a level 0.5 g per kg of concentrate and that of group 3 with a cornus extract plus oregano and thyme essential oils, at a level of 0.5 g cornus extract, 0.01 g oregano and 0.005 g thyme essential oil per kg of concentrate, respectively (Cornus and CorOrThym groups). All plant extracts were provided by ORIZON Diatrofiki, Serres, Greece. At the start of the trial, the ewes were in the third month of their second lactation (on average two months after lamb weaning) and were acclimatized to the experimental conditions for two weeks prior to starting. The mean initial BW was 54.4 ± 1.65 kg and the trial lasted for 45 days plus acclimatization of two weeks, and the mean milk yield was 1.73 L/d. The final BW was 55.3 ± 1.44 kg. After weaning, the ewes were machine milked twice daily (at 6:30 and 17:00 h) in a 24-parallel milking parlor with individual ear-tag identification.

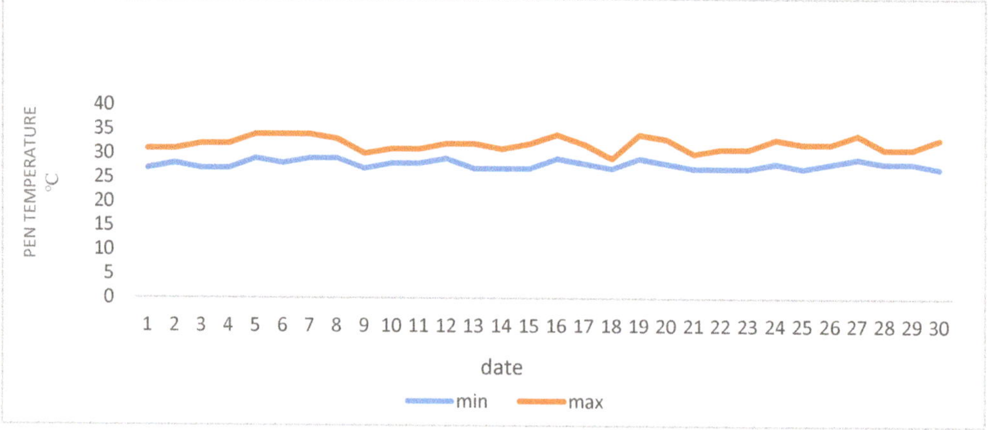

Figure 1. Minimum and maximum temperatures recorded on the sheep farm in July 2020.

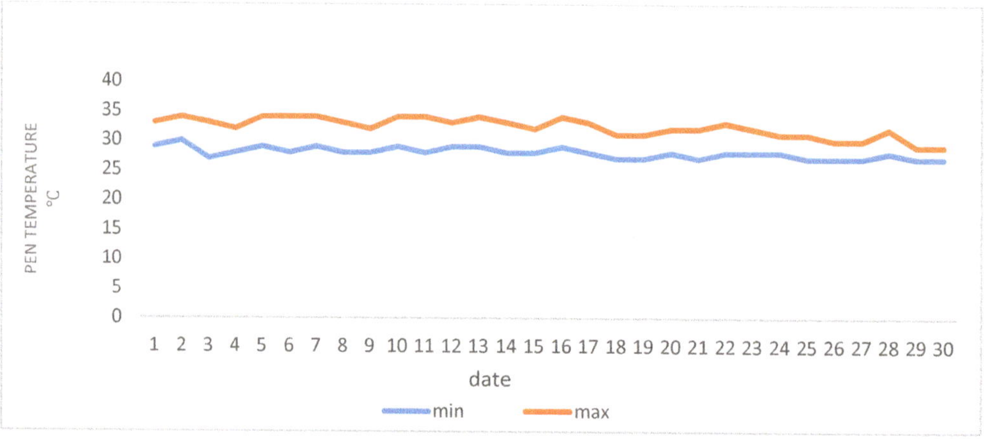

Figure 2. Minimum and maximum temperatures recorded on the sheep farm in August 2020.

All ewes were fed the same total diet, which was offered twice daily in equal quantities. The daily forage allowance per ewe was 1.5 kg of Lucerne hay and 0.30 kg of wheat straw on a fresh weight basis. The concentrate feed consisted of corn grains, barley grains, wheat bran, extracted soybean meal, sunflower cake and mineral and vitamin premix (Table 1). The concentrates were fed at the level of 1.51 kg per ewe per day; thus, the total feed allowance was 3.31 kg fresh food/ewe/day. The feed allowance was kept constant throughout the experimental period, according to commercial practices. Table 1 shows the chemical analysis of the bilateral ration that was conducted according to AOAC [28]. Feed samples were collected monthly and subsequently analyzed for each group. Any orts were collected and taken into account in the calculations. Fresh water was available ad libitum.

Table 1. Daily ingredient allowance and chemical composition of the diet offered to dairy ewes [1].

Ingredients (on Fresh Weight Basis)	Diet (g/day/ewe)
Lucerne Hay	1500
Wheat Straw	300
Corn	900
Wheat Bran	50
Sunflower Seed Meal-36	150
Soya Bean Meal-47	350
Premix [1] of Vitamins and Inorganic Minerals	60
Chemical Analysis (%)	-
Dry Matter	88.1
Crude Protein (N × 6.25)	17.3
Ether Extract	3.1
Crude Fiber	16.2
NDF	25.4
ADF	16.7
ADL	4.5
Ash	4.6
Starch	19.3
Sugars	6.7
Calculated Analysis	-
Calcium	16.5 g/kg
Phosphorus (total)	7.1 g/kg
PDI	102 g/kg
PDIA	58 g/kg
UF_L	0.71 g/kg

[1] Vitamin and mineral mix contained (per kg dry matter (DM) of concentrate): 8000 IU of vitamin A; 90 mg of vitamin E; 3000 IU of vitamin D_3; 1.5 mg/kg biotin; 6 mg/kg niacin; 45 mg/kg choline; 0.2 mg Co; 3 mg I; 100 mg/kg Fe; 50 mg Mn; 0,45 mg Se; 150 mg Zn; 6 g of NaCl; 4 g of sulphur; 10 g of magnesium oxide; 15 g of monocalcium phosphate and 21 g of limestone. NDF = neutral detergent fiber, ADF = acid detergent fiber, ADL = acid detergent lignin, PDI = protein digestible in the small intestine, PDIA = protein digestible in the small intestine supplied by rumen–undegraded dietary protein, UF_L = forage unit for lactation [29].

2.3. *Chemicals*

All solvents or analytical standards such as 2,2-diphenyl-picrylhydrazyl (DPPH) stable radical, Trolox™, Folin–Ciocalteu reagent, gallic acid, nordihydroguaiaretic acid (NDGA), dimethyl sulfoxide (DMSO), trichloroacetic acid (TCA) and sodium chloride (NaCl), were purchased from Sigma-Aldrich, Chemie GmbH (Taufkirchen, Germany). For gas chromatography (GC), helium gas was purchased from Afoi Thomadaki (Thessaloniki, Greece).

2.4. *Characterization of Cornelian Cherry Extract (Cornus mas* L.*)*

2.4.1. Determination of the Antiradical Activity (A_{AR})

For the antiradical activity (A_{AR}) determination, a previously described protocol was used [30] with slight modifications. In brief, an aliquot of 0.025 mL of sample was added to 0.975 mL DPPH solution (100 μM in MeOH) and the absorbance was read at t = 0 and t = 30 min. Trolox™ equivalents (mM TRE) were determined based on linear regression, after plotting $\%\Delta A_{515}$ against the respective known concentration of a Trolox™ solutions, where

$$\%\Delta A = \frac{A_{515}^{t=0} - A_{515}^{t=30}}{A_{515}^{t=0}} \times 100 \quad (1)$$

expressing the difference in antiradical activity, where $A_{515}^{t=30}$ is the absorbance of the sample at 515 nm after the necessary time to reach the plateau (30 min) and $A_{515}^{t=0}$ is the absorbance of the DPPH solution at 515 nm.

Results were expressed as μmol TRE/g of cornelian cherry.

2.4.2. Determination of Total Polyphenol Yield (YTP)

The total phenolic content of the extracts was determined by using the Folin–Ciocalteu method [30]. The yield in total polyphenols (Y_{TP}) was expressed as mg gallic acid equivalents (GAE) per 100 g of cornelian cherry extract after 180 min of incubation in dark.

2.4.3. Determination of Total Monomeric Anthocyanins

The total monomeric anthocyanin content of the extract was determined according to a protocol using the pH-differential method [31,32]. Briefly, the total monomeric anthocyanin content was determined by measuring the absorbance at 510 nm and 700 nm, against distilled water. Measurements were made after dilution of the extracts in buffer solutions of pH 1.0 and pH 4.5, accordingly. The calculation was based on two equations:

$$A = (A_{\lambda max} - A_{700})_{pH1.0} - (A_{\lambda max} - A_{700})_{pH4.5} \quad (2)$$

where $A_{\lambda max}$ is the absorbance of the sample at 510 nm.

$$\text{Total monomeric anthocyannins } (\text{mg}/100\text{ g}) = A * M_w * Df * 1000/_{\varepsilon * 1} \quad (3)$$

where M_W = molecular weight of cyanidin-3-glucoside (449.2 g/mol); Df = dilution factor; l = path length of the cuvette in cm; ε = molar extinction coefficient of cyanidin-3-glucoside (26,900 L/mol/cm); 1000 = conversion of g to mg. The results were expressed as mg cyanidin-3-glucoside per 100 g of cornelian cherry dry matter. All analyses were done in triplicate.

2.5. Essential Oil Analysis

Essential oil analyses were performed on a Shimadzu GC-2010-GCMS-QP2010 (Kyoto, Japan) system operating at 70 eV. The gas chromatograph (GC) was equipped with a split/splitless injector (230 °C) and a fused silica column INNOWAX (Santa Clara, CA, USA) (30 m × 0.25 mm, film thickness: 0.25 μm). The temperature program was from 50 °C (20 min) to 250 °C, at a rate of 3 °C/min. Helium was used as a carrier gas at a flow rate of 1.0 mL/min. The injection volume of each sample was 1 μL. The injector was set at 230 °C and operated in split mode (split ratio = 1:10), whereas the GC–MS transfer line and the ion source were set at 300 °C and 230 °C, respectively. The mass spectrometer was operated in electron ionization mode (70 eV) and full-scan mass spectra were acquired from m/z 100 to 600. The relative percentage amounts of the separated compounds were calculated from the total ion chromatogram by a computerized integrator. The identification of the components was based on comparison of their mass spectra with those of NIST21 and NIST107 [33] and by comparison with literature data [34]. Essential oils were often subjected to co-chromatography with authentic compounds.

2.6. Determination of Total Phenolic Content of Experimental Diets, Milk, Yoghurt and Cheese

The concentrate fed to the three different experimental groups was also analyzed for its total phenolic content according to the method of Singleton et al. (1999) [35] using the Folin–Ciocalteu assay and expressed as gallic acid equivalents (GAE). An aliquot (1 mL) of the sample or a standard solution of gallic acid (blank, 100, 200, 300, 400 and 500 μg/mL) was transferred into a 25 mL volumetric flask, containing 9 mL of distilled water. Subsequently, 1 mL of Folin–Ciocalteu phenol reagent (Merck, Darmstadt, Germany) was added to the mixture and shaken. After 5 min, 10 mL of a sodium carbonate (Na_2CO_3) solution (7% *w/w*) was added to the mixture. After incubating for 90 min at room temperature, the absorbance was measured against the reagent blank at 550 nm with a UV-VIS spectrophotometer (UV-1700 PharmaSpec, Shimadzu, Kyoto, Japan). For the milk, yoghurt and cheese samples, 5 g were mixed with 25 mL of an aqueous methanol solution (ratio of methanol to water: 75:15). The mixture was stirred for 30 min and centrifuged for 15 min at 7200 rpm. The upper phase was filtered through a nylon syringe filter (filter diameter: 25 mm, pore size:

0.22 μm, Frisenette, Knebel, Denmark), and analyzed for gallic acid equivalents using the procedure described above.

2.7. Feed and Essential Oils Interaction with DPPH

The antioxidant activity of the essential oils and the corresponding feed stuff was determined with respect to hydrogen-donating or radical-scavenging ability, according to the method of Peperidou et al. (2014) [36], using the stable radical 1.1-diphenyl-2-picrylhydrazyl (DPPH). Ten microliters of the test samples, which were dissolved in methanol (20 mg/mL stock solution), were added to a solution of DPPH (0.1 mM in DMSO). After 20 and 60 min the absorbance was recorded at 517 nm and the percentage of reducing activity (RA) was calculated and compared with the reference compound, nordihydroguaiaretic acid (NDGA).

2.8. Determination of Milk Yield and Composition

The milk yield of each ewe was recorded weekly by the total volume measurement method. Each sheep was recognized by their individual ear tag. Individual milk samples were analyzed for total fat, total protein, lactose, total solids, total viable counts and solid-not-fat (SNF) by means of near-infrared spectroscopy using a MilkoScan 4000 (FOSS Electric, Integrated Milk TestingTM, Hillerød, Denmark) and somatic cell counts were determined using a Fossomatic 5000 Basic (FOSS Electric, Denmark). Milk acidity was measured using a portable titrator.

2.9. Preparation of Yoghurt and Determination of Bacteria Culture Viability

Yoghurt was produced by sheep milk. Five liters of milk from each group (Control, Cornus and CorOrThym) were collected and heated for 8 min at 92 °C, followed by cooling to 42 °C, and the addition of the standard yogurt starter culture under aseptic conditions. Subsequently, the milk was transferred into 200-mL plastic retail containers, sealed, kept quiescent and incubated at 42 °C for 3 h. Afterwards, the samples were immediately stored at 5 °C. After 72 h, the yoghurt samples were subjected to physicochemical and organoleptic analyses. All yoghurt trials were repeated twice. Enumeration of *Lactobacilus bulgaricus* and *Streprococcus thermophilus* populations were conducted using the pour plate technique and carried out after 10 days of storage at 4 °C. An aliquot of 1 g of yoghurt was diluted in 9 mL of peptone water, containing sodium chloride (0.9% *w/v*), followed by seven sequential dilutions (10^{-1}–10^{-7}). A quantity of 1 mL of the diluted sample of 10^{-4} up to 10^{-7} was placed in M17 agar (Merck, Germany) and each petri was incubated at 37 °C for 72 h. For each sample, the mean numbers of colony forming units (CFU) per gram of yoghurt was estimated.

2.10. Feta Cheese Production and Determination of its Physico-chemical Characteristics

Fresh sheep milk was standardized to a protein-to-fat ratio of 1.0, pasteurized at 63 °C for 30 min, and then cooled to 35 °C. Cheese manufacturing was carried out according to the traditional procedure. In brief, calcium chloride solution (0.2% *w*/*v*) was added to the cheese milk. The milk was held for 30 min at 35 °C for culture maturation. Afterwards, powdered calf rennet was added to achieve coagulation in about 50 min at 35 °C. After coagulation, the curd was cut into cubes of 2 cm and left to rest for 10 min. The sliced curd was then transferred into molds, stored at 18 °C and turned every hour for 3 h for whey drainage, surface salted using dry salt and left overnight to complete whey drainage. Following, the cheeses were placed in containers with brine (7%) and kept at 18 °C for approximately 10 days for cheese ripening. Subsequently, the cheese containers were sealed and stored at 4 °C for a total ripening period of 60 days. Samples were collected on day 60 of ripening and subjected to physicochemical analyses. The pH of the cheeses was determined by a pH meter (GLP-21, CRISON Instruments SA., Barcelona, Spain). The moisture content was determined by the sea sand method, e.g., by thoroughly mixing 2 g of sample with 20 g sea sand and heating at 105 °C until it was constant in weight [37].

The sodium chloride (NaCl) content was determined according to the standard method of the International Dairy Federation [38] and expressed as salt-in-moisture concentration. Finally, total nitrogen (TN) was determined using the Kjeldahl method [28] and the fat content was determined using the Gerber van Gulik method [39].

2.11. Determination of the Fatty Acid Composition in Milk, Yoghurt and Feta cheese

Fatty acid methyl esters (FAMEs) were prepared according to [40] Bligh and Dyer (1959) method and [41] the International Organisation for Standardisation (ISO) (2002) as reported previously by Papaloukas et al., 2016 [42]. Analysis of FAMEs was performed on an Agilent Technologies 6890N gas chromatograph (GC), equipped with a flame ionization detector (FID) (Santa Clara, CA, USA) and a DB-23 capillary column (60 m × 0.25 mm i.d., 0.25-μm film thickness). Each peak was identified and quantified using a 37 component FAME mix (Supelco, 47885-U) and the PUFA-2, Animal source (Supelco, 47015-U) (Sigma-Aldrich, Taufkirchen, Germany). After analysis, the saturated fatty acids (SFA), monounsaturated fatty acids (MUFA), polyunsaturated fatty acids (PUFA), unsaturated fatty acids (UFA), n-3 PUFA and n-6 PUFA were further grouped together. The index for atherogenicity (AI) was determined as suggested by [43] Ulbricht and Southgate (1991), and the $\Delta-9$ desaturase activity indexes were calculated using the following four ratios: C14:1/C14:0, C16:1/C16:0, C18:1/C18:0 and CLA/VA.

2.12. Determination of TBARS and Protein Carbonyls in Milk, Yoghurt and Feta Cheese

Fresh milk samples were collected on days 1, 21 and 42 of the trial. The samples were analyzed immediately. For the determination of TBARS (thiobarbituric acid reactive substances), an aliquot of 100 μL of the fresh milk sample was mixed with 500 μL of trichloroacetic acid (TCA) solution (35 % *w/v*) and 500 μL of Tris–HCl (200 mmol/L; pH 7.4), and incubated for 10 min at 20 °C. Subsequently, 1 mL of 2MNa_2SO_4 and 55 mM thiobarbituric acid solution were added, and the samples were incubated at 95 °C for 45 min. Finally, the absorbance was measured at 530 nm against a blank sample using an UV-VIS spectrophotometer (UV-1700 PharmaSpec, Shimadzu, Japan). Results were expressed as μmol MDA per liter of milk.

Protein carbonyl determination was based on the method of Patsoukis et al. (2004) [44]. In particular, 50 μL of a TCA solution (20% *w/v*) was added to 50 μL of sample homogenate (diluted 1:2 v/v). The mixture was then kept in an ice bath for 15 min and centrifuged at 15,000× *g* for 5 min at 4 °C. The supernatant was discarded. To the remaining pellet, 500 μL of a 2,4-dinitrophenylhydrazine (DNPH) solution (10 mmol/L in 2.5 N HCl) were added. The blank was prepared by addition of 500 μL of 2.5 N HCl to the pellet. In this assay, carbonyl formation is detected by the reaction of protein carbonyls with 2,4-dinitrophenylhydrazine (DNPH) and its subsequent conversion to 2,4-dinitrophenylhydrazone (DNP-hydrazone), which was measured at 375 nm. Calculation of protein carbonyl concentration was based on the molar extinction coefficient of DNPH (22,000 M^{-1} cm^{-1}).

2.13. Determination of Blood Serum Parameters

The influence of the different diets on animal health indices was also evaluated by analyzing specific biochemical parameters in the blood serum. Blood samples were obtained from each ewe by jugular venipuncture into 10 mL vacuum tubes without anticoagulant with a needle, on the first and last day of the experiment. Samples were centrifuged (1600× *g* for 15 min in 4 °C), and the collected serum was frozen at −20 °C until analysis. The biochemical parameters tested in the serum were total proteins, albumins, glucose, blood urea nitrogen (BUN), creatine and gamma-glutamyl transpeptidase (γ-Gt) and they were assayed using an automated analyzer (TARGA CLIN/CHEM Analyzer, BT 1500 Biotechnica instruments Roma, Italy).

2.14. Statistical Analysis

Data on total milk production and composition were analyzed by ANOVA in the general linear model of the SPSS 25.00 statistical package (SPSS Inc., Chicago, IL). Individual ewes were considered replicates nested within pens; the pen was the statistical unit for the analysis. Data on total milk yield and milk, cheese, yoghurt composition, serum biochemical parameters were analyzed by means of one-way ANOVA. The homogeneity of the variances was tested using the Levene test. The Tukey multiple comparison test was carried out to assess any significant differences at a probability level of $p < 0.05$ between the experimental treatments, when a significant effect of treatment was detected by means of the ANOVA.

3. Results

All the animals consumed all their offered daily food allowance of 3.31 kg. Total observed feed refusals were estimated to be 0.05–1% of the total food allowance. None of the animals showed symptoms of any health affliction. The diet of the control group contained 0.18 mg GAE/g dry matter (DM). The cornus-supplemented diet contained 0.50 mg GAE/g DM; whereas the CorOrThym diet contained 0.66 mg GAE/g DM. The results of the photometric analysis of the cornelian extracts are presented in Table 2. In Table 3 the analyses of the oregano and thyme essential oils are presented. Table 4 presents the results relating to the DPPH and TP antioxidant activity of the diets and oregano and thyme essential oils.

Table 2. Photometric analysis of Cornelian cherry extract.

Composition	Extract
Total phenolics (mg/100 g) *	7611 ± 623
Antiradical activity (µmol TRE/g) **	58,057 ± 741
Total monomeric anthocyanins (mg/100 g) ***	382 ± 15

* as gallic acid. ** TRE = Trolox equivalent. *** cyanidin-3-glucosid.

Table 3. Composition (%) of oregano and thyme essential oil.

Compounds [a]	%	Compounds [a]	%
Oregano	-	**Thyme**	-
α-Pinene	0.77	α-Pinene	1.84
α-Thujene	0.52	α-Thujene	0.94
Camphene	0.07	Camphene	1.03
β-Pinene	0.09	β-Pinene	0.36
α-Phellandrene	0.14	δ 3-Carene	0.11
β-Myrcene	1.26	α-Phellandrene	0.21
α-Terpinene	1.02	β-Myrcene	1.87
D-Limonene	0.17	α-Terpinene	2.07
γ-Terpinene	7.57	D-Limonene	0.68
3-Octanone	0.27	Eucalyptol	1.64
p-Cymene	7.29	γ-Terpinene	12.55
3-Octanol	0.08	p-Cymene	38.34
1-Octen-3-ol	1.01	Terpinolene	0.18
cis-Sabinenehydrate	0.68	3-Octanol	0.20
trans-Sabinenehydrate	0.34	1-Octen-3-ol	1.19
Linalool	0.48	*cis*-Sabinene Hydrate	1.05
4-Octanol	0.05	Camphor	0.43
Bornyl Acetate	0.06	*trans*-Sabinene Hydrate	0.30
β-Caryophyllene	2.24	Linalool	3.63
Thymol Methyl Ether	0.06	Bornyl Acetate	0.17
Dihydrocarvone	0.09	β-Caryophyllene	1.64
1-Terpinen-4-ol	0.94	Thymol Methyl Ether	0.85

Table 3. *Cont.*

Compounds [a]	%	Compounds [a]	%
Oregano	-	**Thyme**	-
Carvacrol Methyl Ether	0.50	1-Terpinen-4-ol	0.48
α-Caryophyllene	0.18	Carvacrol Methyl Ether	0.63
Borneol	0.95	Borneol	1.72
β-Bisabolene	0.87	δ-Cadinene	0.21
Germacrene D	0.15	Caryophyllene Oxide	0.13
Caryophyllene oxide	0.14	Thymol	20.32
Thymol	13.51	Hinesol	0.14
Hinesol	0.33	Carvacrol	2.52
Carvacrol	57.23	Apiol	0.16
Apiol	0.41	Ledol	0.04

[a] Compounds are listed in order of elution from an INNOWAX capillary column.

Table 4. Photometric analysis of oregano and thyme essential oils and sheep diets.

Sample [2]	DPPH (20 mg/mL)		TP *
	20 min	60 min	GA
Oregano essential oil (EO)	87.70	93.55	837.93
Thyme EO	24.15	27.87	535.0
Feed Control	46.23	44.56	18.02
Feed Cornus	75.27	85.87	50.25
Feed CorOrThym	62.36	69.56	66.32
NDGA [1]	81	93	

[1] Results are given as means of groups (n = 5 = subgroups); [2] Control, Cornus and CorOrThym represent diets supplemented with corresponding plant extracts; * TP = total phenolics as gallic acid (GA) equivalents (mg/L); [1] NDGA = nordihydroguaiaretic acid.

3.1. Milk Yield

Milk yield followed the typical lactation curve and no interactions were observed between treatments and time (trial duration) on daily milk yield over the tested period. However, supplementation of the diet with cornus extract, with and without essential oil, seemed to improve the daily milk yield compared to the control group. The average milk yield in the first days of the trial did not differ among the groups and was found to be 1.74 L/d for the control group and 1.72 L/d for both the Cornus and CorOrThym group, respectively. On day 21 a significant increase in average milk yield of 1.76 L/d (Cornus group) and 1.92 L/d (CorOrThym group), compared to an average milk yield of 1.72 L/d for the control group, was noticed (Table 5). This pattern remained unchanged on the last day of the experiment. However, there was no connection between cornus extract and EO addition on milk composition, except for urea and somatic cell counts (SCCs), which were found to be lower in the more highly EO-supplemented groups (Table 6). Although we did not measure any changes in body composition, occasional observations did not suggest that the effects of the EO supplementation were achieved through changes in body mobilization between the treatments. In addition, BW was similar among the groups at the start and at the end of the trial.

Table 5. Effects of diet supplementation with cornus extract and oregano and thyme essential oils on milk yield.

Milk Yield [1]	Control [2]	Cornus	CorOrThym	SEM [3]	*p*-Value
Day 0	1.74	1.72	1.72	0.03	NS
Day 21	1.72 ^{b}	1.76 ^{ab}	1.92 ^{a}	0.04	0.04
Day 42	1.71 ^{b}	1.78 ^{ab}	1.92 ^{a}	0.03	0.04

[1] Results are given as means of groups (n = 5 = subgroups); [2] Control, Cornus and CorOrThym represent groups of ewes fed a basal diet supplemented with corresponding plant extracts; ^{a,b} values with a superscript in common in the same line do not differ significantly. [3] SEM = standard error of the mean. NS = not significant.

Table 6. Effects of diet supplementation with cornus extract and oregano and thyme essential oils on milk composition and somatic cell counts (SCCs) of ewes during lactation.

Composition [1]	Control [2]	Cornus	CorOrThym	SEM [4]	*p*-Value
		Day 0			
Protein (%)	5.65	5.61	5.62	0.26	NS
Fat (%)	5.85	5.88	5.84	0.35	NS
Lactose (%)	4.77	4.77	4.78	0.22	NS
Ash (%)	0.85	0.89	0.86	0.56	NS
SNF [3] (%)	11.24	11.21	11.23	0.38	NS
SCC ($\times 10^3$/mL)	26	24.0	27	25.6	NS
pH	6.72	6.71	6.71	0.05	NS
Acidity	22.1	22.2	22.5	0.36	NS
Urea (mg/100 mL)	58.4	52.1	59.3	2.95	NS
Total viable counts ($\times 10^3$/mL)	160	160	160	15.3	NS
		Day 42			
Protein (%)	5.85	5.89	5.95	0.25	NS
Fat (%)	5.95 ^{b}	6.16 ^{b}	6.39 ^{a}	0.39	0.04
Lactose (%)	4.82	4.82	4.88	0.26	NS
Ash (%)	0.85	0.89	0.86	0.55	NS
SNF [3] (%)	11.31	11.32	11.31	0.61	NS
SCC ($\times 10^3$/mL)	36 ^{a}	33 ^{b}	32 ^{b}	22.4	0.01
pH	6.72	6.71	6.71	0.04	NS
Acidity	22.1	22.2	22.5	0.32	NS
Urea (mg/100 mL)	48.3 ^{a}	32.1 ^{b}	29.4 ^{b}	2.08	0.004
Total viable counts ($\times 10^3$/mL)	200 ^{a}	140 ^{b}	130 ^{b}	13.5	0.005

[1] Results are given as means of groups (n = 5 = subgroups); [2] Control, Cornus and CorOrThym represent groups of ewes fed basal diet supplemented with corresponding plant extracts; [3] SNF = solids-not-fat content of milk. SCC =somatic cell counts, ^{a,b} values with a superscript in common in same line do not differ significantly. [4] SEM = standard error of the mean. NS = not significant.

3.2. *Milk, Yoghurt and Feta: Composition, Fatty Acid Composition and Oxidation Status*

The composition of the milk from the three investigated groups is presented in Table 6. Yoghurt and feta cheese composition was within the normal standards and no major differences were noted. The acceptance of the products was satisfactory in regard to taste and aroma.

Regarding the levels of the main fatty acids in milk, yoghurt and cheese—namely, palmitic and linoleic acids—no major differences were noted (Tables 7 and 8).

Table 7. Effects of diet supplementation with cornus extract and oregano and thyme essential oils on fatty acid profile of milk (% of total identified fatty acids).

Fatty Acids [1]	Control [2]	Cornus	CorOrThym	SEM [4]	*p*-Value
C4:0	1.01	1.00	0.91	0.03	NS
C6:0	1.34	1.33	1.37	0.07	NS
C8:0	2.11	2.03	2.13	0.14	NS
C10:0	9.60	9.36	9.48	0.65	NS
C11:0	0.37	0.37	0.39	0.03	NS
C12:0	7.08	7.43	6.96	0.50	NS
C13:0	0.11	0.10	0.10	0.01	NS
C14:0	15.84	18.01	15.96	1.25	NS
C14:1	0.52	0.74	0.60	0.06	NS
C15:0	0.98	0.85	0.91	0.05	NS
C15:1	0.22	0.16	0.17	0.01	NS
C16:0	29.44	31.02	29.07	1.74	NS
C16:1	0.85	1.21	1.17	0.10	NS
C17:0	0.52	0.39	0.50	0.04	NS
C17:1	0.22	0.13	0.22	0.01	NS
C18:0	8.69	6.08	7.08	1.17	NS
18:1 trans-11	0.33	0.19	0.25	0.07	NS
C18:1 n-7 cis-VA	0.59	0.48	0.59	0.31	NS
18:1cis cis-9	15.87	15.23	17.16	1.75	NS
18:2 n-6 trans	0.62	0.69	0.88	0.34	NS
18:2 n-6 cis	2.27	1.91	2.22	0.18	NS
18:3n6	0.11	0.12	0.16	0.02	NS
18:3n3	0.80	0.57	0.88	0.16	NS
CLAcis-9, trans-11	0.35	0.48	0.52	0.12	NS
C20:0	0.03	0.00	0.12	0.01	NS
C21	0.02	0.05	0.07	0.02	NS
C20:2	0.11	0.08	0.12	0.02	NS
SCFA (C4-C11 sat)	14.43	14.09	14.29	0.86	NS
MCFA (C12-C16 sat)	53.45	57.41	53.00	3.40	NS
LCFA (C17-C21 sat)	9.26	6.52	7.76	1.26	NS
PUFA	4.26	3.84	4.78	0.81	NS
MUFA	18.60	18.14	20.17	2.05	NS
UFA	22.86	21.99	24.95	2.80	NS
SFA	77.14	78.02	75.05	2.79	NS
*AI((C12:0 + 4 * C14:0 + C16:0)/UFA)*	4.37	5.03	4.00	0.48	NS
C14:1/C14:0 [3]	0.03	0.04	0.04	0.01	NS
C16:1/C16:0	0.03	0.04	0.04	0.01	NS
C18:1/C18:0	1.86	2.54	2.46	0.01	NS
C18:2cis9, trans-11/C18:1 trans11	0.60	0.99	0.88	0.07	NS
SFA/UFA	3.38	3.55	3.01	0.30	NS

SCFA: short chain fatty acids; MCFA: medium chain fatty acids; LCFA: long chain fatty acids; PUFA: polyunsaturated fatty acids; MUFA: monounsaturated fatty acids; SFA: saturated fatty acids; UFA: unsaturated fatty acids; [1] results are given as means of groups ($n = 5$ = subgroups); [2] Control, Cornus and CorOrThym represent groups of ewes fed a basal diet supplemented with corresponding plant extracts; [3] the $\Delta-9$ desaturase activity indexes were calculated using the following four ratios: C14:1/C14:0, C16:1/C16:0, C18:1/C18:0 and C18:2cis9,trans-11/C18:1 trans11; [4] SEM = standard error of the mean. NS = not significant.

Table 8. Effects of diet supplementation with cornus extract and oregano and thyme essential oils on fatty acid profile of yoghurt and Feta cheese (% of total identified fatty acids).

Fatty Acids [1]	Control [2]	Cornus	CorOrThym	Control	Cornus	CorOrThym	SEM [4]	*p*-Value
-		Yoghurt			Feta Cheese		-	-
C4:0	0.86	1.14	1.37	0.77	0.85	0.76	0.23	NS
C6:0	1.04	1.27	1.32	1.30	1.43	1.27	0.07	NS
C8:0	1.67	1.97	2.03	1.78	2.32	2.06	0.09	NS

Table 8. *Cont*

Fatty Acids [1]	Control [2]	Cornus	CorOrThym	Control	Cornus	CorOrThym	SEM [4]	*p*-Value
C10:0	7.12	8.52	8.31	7.76	9.81	8.70	0.39	NS
C11:0	0.29	0.31	0.34	0.32	0.39	0.22	0.01	NS
C12:0	5.72	6.49	6.02	5.70	7.17	6.72	0.25	NS
C13:0	0.14	0.13	0.14	0.12	0.15	0.13	0.00	NS
C14:0	14.70	15.42	14.30	15.27	15.69	14.76	0.35	NS
C14:1	0.45	0.32	0.39	0.44	0.32	0.38	0.03	NS
C15:0	1.40	1.03	1.16	1.19	0.95	1.05	0.04	NS
C15:1	0.26	0.16	0.18	0.27	0.18	0.19	0.01	NS
C16:0	30.92	28.98	27.52	32.02	27.55	28.12	0.84	NS
C16:1	1.41	0.63	1.33	1.42	0.98	1.06	0.13	NS
C17:0	0.65	0.50	0.44	0.66	0.52	0.58	0.02	NS
C17:1	0.29	0.17	0.22	0.26	0.20	0.24	0.01	NS
C18:0	7.83	8.33	7.87	7.28	7.92	8.36	0.59	NS
18:1 trans-11	0.33	0.32	0.54	0.36	0.36	0.51	0.04	NS
C18:1 n-7 cis-VA	0.77	0.64	1.37	0.94	0.74	1.16	0.13	NS
18:1 *cis-9*	18.90	18.77	18.50	16.79	17.43	17.69	0.66	NS
18:2 n-6 trans	0.75	0.62	1.16	0.62	0.66	0.85	0.17	NS
18:2 n-6 cis	2.45	2.53	3.14	2.36	2.70	2.71	0.16	NS
18:3n6	0.29	0.10	0.16	0.20	0.22	0.28	0.03	NS
18:3n3	0.82	0.76	0.95	1.02	0.71	0.85	0.19	NS
CLAcis-9, trans-11	0.66	0.67	0.82	0.80	0.53	0.86	0.05	NS
C20:0	0.08	0.11	0.18	0.09	0.08	0.18	0.10	NS
C21	0.09	0.08	0.11	0.09	0.05	0.17	0.02	NS
C20:2	0.13	0.01	0.15	0.20	0.11	0.16	0.03	NS
SCFA	10.98	13.21	13.36	11.92	14.79	13.00	0.62	NS
MCFA	52.88	52.04	49.15	54.29	51.51	50.78	1.34	NS
LCFA	8.63	9.03	8.59	8.11	8.57	9.29	0.61	NS
PUFA	5.09	4.70	6.38	5.20	4.93	5.71	0.45	NS
MUFA	22.41	21.02	22.53	20.48	20.21	21.22	0.74	NS
UFA	27.50	25.72	28.91	25.68	25.14	26.93	1.13	NS
SFA	72.50	74.28	71.10	74.32	74.86	73.07	1.13	NS
*AI(C12:0 + 4 * C14:0 + C16:0)/UFA)*	3.47	3.78	3.14	3.85	3.88	3.49	0.20	NS
C14:1/C14:0 [3]	0.03	0.02	0.03	0.03	0.02	0.03	0.01	NS
C16:1/C16:0	0.05	0.02	0.05	0.04	0.04	0.04	0.01	NS
C18:1/C18:0	2.46	2.29	2.42	2.36	2.25	2.18	0.12	NS
C18:2cis9, trans-11/C18:1 trans11	0.85	1.05	0.60	0.85	0.71	0.74	0.01	NS
SFA/UFA	2.64	2.89	2.46	2.89	2.98	2.71	0.01	NS

SCFA: short chain fatty acids; MCFA: medium chain fatty acids; LCFA: long chain fatty acids; PUFA: polyunsaturated fatty acids; MUFA: monounsaturated fatty acids; SFA: saturated fatty acids; UFA: unsaturated fatty acids; [1] results are given as means of groups (n = 5 = subgroups); [2] Control, Cornus and CorOrThym represent groups of ewes fed a basal diet supplemented with corresponding plant extracts; [3] the $\Delta-9$ desaturase activity indexes were calculated using the following four ratios: C14:1/C14:0, C16:1/C16:0, C18:1/C18:0 and C18:2cis9,trans-11/C18:1 trans11; [4] SEM = standard error of mean. NS = not significant.

However, it is noteworthy that the addition of Cornus and CorOrThym to the animals' diet affected the starter culture (*Streptococcus thermophilus*) in the produced yoghurt samples; a reduction of about 2 logs of Streptococcus thermophilus was observed, compared to the control sample (Table 9). TBARS values, gallic acid equivalents and protein carbonyl levels were improved for the group fed with the cornus extract and oregano and thyme oils compared to the control group (Table 10).

Table 9. Effects of diet supplementation with cornus extract and oregano and thyme essential oils on yoghurt bacteria populations.

Yoghurt [1]	Control [2]	Cornus	CorOrThym	SEM [3]	*p*-Value
Streptococcus thermophilus	1.3×10^{9} [a]	2.6×10^{7} [b]	3.9×10^{7} [b]	0.10	0.011
Lactobacillus bulgaricus	ND [4]	ND	ND	-	-

[1] Results are given as means of groups (n = 5 = subgroups); [2] Control, Cornus and CorOrThym represent groups of ewes fed a basal diet supplemented with corresponding plant extracts; [a,b] values with a superscript in common in same line do not differ significantly; [3] SEM = standard error of the mean; [4] ND = not detected.

Table 10. Effects of diet supplementation with cornus extract and oregano and thyme essential oils on antioxidant status of milk, yoghurt and Feta cheese.

Parameter [1]	Control [2]	Cornus	CorOrThym	SEM [3]	*p*-Value
		Milk			
TBARS (ng/mL)	0.14	0.14	0.13	0.17	NS
Protein carbonyls (ng/mL)	23.63 [a]	20.90 [a]	11.81 [b]	0.68	0.02
TP *	0.05 [b]	0.05 [a]	0.58 [a]	0.06	0.01
		Yoghurt			
TBARS (ng/mL)	0.18	0.14	0.14	0.48	0.08
Protein carbonyls (ng/mL)	11.81 [a]	9.54 [a]	4.09 [b]	0.68	0.01
TP *	0.03 [b]	0.08 [a]	0.08 [a]	0.05	0.01
		Feta Cheese			
TBARS (ng/mL)	2.30	0.53	0.48	0.48	0.02
Protein carbonyls (ng/mL)	30.90 [a]	10.90 [b]	7.72 [b]	0.68	0.005
TP *	0.02 [c]	0.09 [b]	0.13 [a]	0.03	0.003

[1] Results are given as means of groups (n = 5 = subgroups); [2] Control, Cornus and CorOrThym represent groups of ewes fed a basal diet supplemented with corresponding plant extracts; * as gallic acid equivalents (ng/mL); [a,b,c] values with a superscript in common in the same line do not differ significantly; [3] SEM = standard error of the mean. NS = not significant.

3.3. Blood Serum Parameters

The results of the blood biochemical parameters are presented in Table 11. Diet supplementation with cornus extract with/without EO did not seem to affect the levels of total urea, creatin and gamma-glutamyl transpeptidase (γ-GT) in the serum among the experimental groups. However, the level of total protein decreased gradually in the Cornus and cornus, oregano and thyme (CorOrThym) groups towards the end of the experimental period. Interestingly, the level of total albumin was increased in the control and Cornus groups but decreased in the CorOrThym group.

Table 11. Effects of diet supplementation with cornus extract and oregano and thyme essential oils on blood parameters [1] of dairy ewes on day 1 and day 42.

Serum [1]	Control [2]	Cornus	CorOrThym	SEM [10]	*p*-Value
		Day 1			
GLU [3] (g/dL)	14.8	12.8	18.8	1.57	NS
TP [4] (g/dL)	10.28	11.22	10.94	0.24	NS
ALB [5] (g/dL)	4.32 [b]	4.62 [b]	5.06 [a]	0.11	0.02
UR [6] (g/dL)	66.4	53.4	66.6	3.42	NS
CR [7] (g/dL)	0.83	0.84	0.76	0.02	NS
γ-GT [8] (g/dL)	53	57.1	37.06	4.39	NS
TBIL [9] (g/dL)	0.79	1.11	1.45	0.21	NS

Table 11. *Cont.*

Serum [1]	Control [2]	Cornus	CorOrThym	SEM [10]	p-Value
		Day 42			
GLU [3] (g/dL)	53.2	40.36	28.4	4.91	NS
TP [4] (g/dL)	11.76	10.44	7.32	0.85	0.08
ALB [5] (g/dL)	5.24	5.9	3.66	0.52	NS
UR [6] (g/dL)	67.6	78.4	65.4	6.74	NS
CR [7] (g/dL)	1.18 ^{ab}	1.42 ^{a}	0.89 ^{b}	0.08	0.01
γ-GT [8] (g/dL)	32.13	22	45.2	8.03	0.07
TBIL [9] (g/dL)	3.79	7.23	1.36	1.06	0.06

[1] Results are given as means of groups (n = 5 = subgroups); [2] Control, Cornus and CorOrThym represent groups of ewes fed a basal diet supplemented with corresponding plant extracts; ^{a,b} values with a superscript in common in the same line do not differ significantly. [3] GLU = glucose content of blood serum; [4] TP = total proteins of blood serum; [5] ALB = albumin content of blood serum; [6] UR = urea of blood serum; [7] CR = creatinine of blood serum; [8] γ-GT = gamma-glutamyltransferase of blood serum; [9] TBIL = total bilirubin of blood serum; [10] SEM = standard error of the mean. NS = not significant.

4. Discussion

Currently, there are several commercially available EO products which are given to ruminants, including dairy animals, in many parts of the world [7,45–47]. In general, essential oils have a positive influence when administrated with dairy sheep rations. Adding 1.25 g/kg DM of thyme essential oil improved both ruminal fermentation and nitrogen metabolism [48]. Another study investigating the effects of oregano essential oils on sheep showed an increase in the population of three primary cellulosic bacteria and ruminal fungi, affecting the rumen fermentation process [49]. However, potential toxic effects of plant essential oils are not scarce and should be taken into consideration before feeding them to animals [50].

The plant extract used in our study included an aqueous cornus extract and a complex of cornus hydro-extract plus oregano and thyme EOs. It was found that the addition of either cornus extract alone or a mixture of cornus extract with oregano and thyme essential oils to the basal ration of dairy sheep improved milk production, milk yield and fat content, whereas differences in total protein, ash and solids-not-fat were not significant. The increasing effect of oregano and thyme on milk yield may be attributed to the galactopoietic effect of the active compounds present in the essential oils [51]. Furthermore, it is suggested that positive effects of galactagogues on milk production may be due to the decrease in circulating biogenic amines such as histamine, tryptamine and tyramine in the blood, which are known to cause the excessive release of catecholamines, leading to a reduction in milk production as well as causing indigestion by inhibiting ruminal mobility and absorption [52]. Other researchers have shown that herbal essential oils can act as galactagogues by enhancing prolactin production and releasing somatotropins, resulting in increased glucose levels in the udder and improved milk production [51–55].

The properties and composition of oregano and thyme have been well investigated. It was reported that their essential oils possess anti-bacterial, anti-fungal, anti-inflammatory and antihistaminic effects [54,55]. Oregano and thyme essential oils increased mammary gland development in rats, followed by a higher milk yield at different stages of lactation [56]. In addition, oregano and thyme essential oils appear to be a potential multipurpose feed additive and may be promising in improving the performance of sheep and lambs [17]. However, there are contradictory opinions in the literature regarding the effect of these plants on milk production; some researchers found that by using oregano and thyme as feed additives, milk yield was increased and that the higher levels remained rather constant [9,57]. On the other hand, it was reported that no positive effects were achieved in ruminants by supplementing their diet with oregano oil or carvacrol, oregano's main constituent. However, it is notable that this study was conducted on dairy cows and the inclusion level of oregano oil or carvacrol was at different levels (50 mg/kg of DM intake) [58]. Additionally, the experimental period lasted for 4 weeks, and no effects were noted on nutrient utilization, ruminal fermentation characteristics, N-excretion, CH_4-production, milk production, milk composition or milk FA profile [58]. In the literature, there are no data about the effect of cornus extracts in combination with oregano or thyme in dairy sheep. Moreover, published studies with EOs and plant extracts have been mainly focused on dairy cows [57,58]. Indeed, dietary hesperidin or naringin improved milk oxidative stability without any further side effects in milk yield, composition, coagulation properties and fatty acid profile in dairy sheep [59], whereas dietary supplementation with plant extracts rich in flavonoids resulted in increased milk yield in cows [60,61].

In contrast to oregano and thyme, few publications have described the effect of dietary cornus supplementation on ruminants. Cornelian cherry (*Cornus mas* L.) is a species of dogwood widely grown in central and south-eastern Europe and Asia. It contains phenolics, including anthocyanins, gallic acid, ellagic acid, quercetin, kaempferol and cyanin [62]. These phenolics have both antioxidant and antimicrobial properties [63,64], whereas ellagic acid also shows immune modulatory activity [64]. Quercetin, another main phenolic compound, was found to reduce inflammation and oxidation damage caused by *Helicobacter pylori* in the mucosa of guinea pigs [65]. Cornus is also rich in ascorbic acid, sugar, organic acids, flavonoids, tannins and other bioactive compounds, among which polyphenols were found in significant amounts [66]. Studies demonstrate that feeding with cornus may reduce the use of antibiotics in young animals post-weaning and the incidence of both diarrhea and mortality in rabbits [67]. Moreover, cornus in beef cattle rations displayed an improvement in DM digestibility, which primarily resulted from the enhanced digestibility of fiber (associated with increased ruminal protozoa counts by feeding cornus) and protein [18]. Another study indicated that increasing dietary cornus was beneficial for the NDF digestibility and also decreased protein degradability, providing a higher concentration of bypass protein. Moreover, in the same study, cornus contributed to the mitigation of acidosis, affecting cattle fed high-grain diets, by altering the nutrient degradability of feeds [19].

In our study, dairy sheep were reared in the hot summer months under semi-temperate climatic conditions. As the mean temperature during the trial period exceeded 28 °C for several hours per day, the animals may have suffered from heat stress. A study by Braun et al. [68] suggested that a blend of plant bioactive essential oils can improve feed efficiency along with calcium homeostasis in dairy cows due to the activation of specific cation-transporting proteins, resulting in an increased uptake of cations like calcium and ammonium [69]. The effects on urea and SCC could be partly an effect of milk dilution. Our findings on milk composition partly agree with a previous study [70], that showed no change in milk concentrations of fat, protein and urea N, and the milk concentrations of fat, protein and lactose content in cows fed up to 2 g/d of a supplement containing essential oil compounds such as thymol, eugenol, vanillin, guaiacol and limonene. In our study we took into account the effect of diet, season and farm management along with parity, BW and stage of lactation in order to avoid confounding the effects of specific plant extracts on milk production and milk quality, as was correctly described [71] when keeping sheep indoors and under standardized feeding conditions without any grazing. Below, we make a further attempt to elucidate the effects of cornus extract and EO supplementation.

The prohibited use of ionophores such as monensin as additives in the EU and the increased interest in 'naturally occurring additives' that can also be used in the 'organic' production of ruminant milk and meat have resulted in a great deal of interest in identifying naturally occurring compounds that positively affect ruminal fermentation and performance [59]. *Thymus vulgaris* supplementation has been found to increase milk yield and the lactation period of Sanjabi ewes [17], as well as improving the milk production of dairy goats [72]. The increase in milk yield and lactation period resulting from thyme supplementation has been associated with the remarkable vasodilator action of thyme essential oil. Particularly, its constituents like bisabolol and bisabolol oxides [72] may act as vasodilators, increasing the flow in the blood vessels which normally supply the mammary gland with all necessary components for milk production, which is helpful during heat stress periods.

Dairy animals are often exposed to stressful agents due to environmental conditions (e.g., very high or very low temperatures), management (high yield, weaning, transportation, feedlot entry) or nutrition (e.g., a high-grain diet). Stressful events have been implicated in promoting oxidative stress through the excessive production of reactive oxygen species or decreased antioxidant defenses [73]. Excessive reactive oxygen species production overwhelms antioxidant defenses, leading to the oxidative damage of biological molecules, disrupting normal metabolism and physiology. However, a reverse relationship between antioxidant intake and stress-related diseases is supported by numerous studies [74,75]. Hence, improving antioxidant capacity through feeding of cornus extract alone or in combination with oregano and thyme is expected to enhance the overall health of sheep and, thus, to improve milk performance. In support of this, a recent study, carried out using 95 control heifers and 95 heifers fed with 0.9 kg/d of cornus extract, showed that the average daily gain increased from 0.89 to 1.12 kg/d [67]. Therefore, it is hypothesized that feeding with red osier dogwood may influence the rumen microflora, immunity (intestine and body) and antioxidant status, and consequently impact feed digestion and immune response. Heat stress is one of the limiting factors affecting the production performance of Mediterranean dairy ruminants, since it evokes a series of drastic changes in animal biological functions, which include a decrease in feed intake efficiency and utilization; disturbances in the metabolism of water, protein, energy and mineral balances; enzymatic reactions; hormonal secretions and blood metabolites [74]. To avoid such situations, proper management approaches are needed. Thus, nutritional techniques can be a part of the solution; in this line, several studies have revealed that supplementing ruminants' rations with various essential oils has shown promising results, such as a decrease in SCC in generally healthy dairy cows or goats under mild heat stress conditions [75–78].

The Chios breed of sheep is a relatively high yielding one [79] and it is mainly found in the Mediterranean basin under intensive indoor conditions that resemble those used for high-yielding dairy cows. Another crucial aim of this study was to evaluate whether a dietary mixture of cornus extract plus oregano and thyme essential oils could enhance their antioxidant properties, showing synergistic properties in simultaneous administration, as it is presented in the literature [80,81], especially during the hot summer period. As a result, diet supplementation with plant extracts and essential oils improved lipid and protein oxidative stability in milk, providing evidence that those animals could better cope with the high ambient temperatures compared to the animals on a normal diet. Moreover, yoghurt and feta cheese made from the milk of animals that were fed the plant extracts presented higher lipid and protein oxidative stability.

Based on the literature, the total anthocyanins of Cornelian cherries range between 106–850 mg cyanidin-3-glycoside/100 g dm and total phenolics between 1070–2696 mg GAE/100 g [82,83]. Cornelian cherries can be considered a good source of anthocyanins, ascorbic acid and phenolics compared to common red berries, and also represent a potential functional food ingredient [82] due to their influence on sensorial properties such as color and astringency. For this reason, analysis of the use of these cherries in foods and beverages has developed during recent decades [83] and it has been reported that total phenolic content, expressed in mg/g, may be almost doubled in a diet supplemented with cornelian extract and essential oils. It was also found that diet supplementation with a combination of cornelian extract and essential oils resulted in the highest interaction with the stable radical DPPH, followed by the cornelian extract itself, which is in agreement with the GC-MS results. Oregano and thyme EOs are rich in carvacrol and thymol, which are phenolic monoterpenoids. It seems that the activity of the essential oil is related to the presence of phenolic compounds. The main role of these components as a reducer of free radicals has been previously reported [84,85].

A great number of studies referring to the addition of essential oils to milk and milk-derived products acknowledge their beneficial effects. It is generally known that the essential oil of thyme has the potential to inhibit various types of pathogenic microorganisms. The addition of thyme and thyme essential oil leads to an increased yoghurt fermentation time, whereas the counts of lactic acid bacteria finally reach the required minimum value for dairy fermented products [86]. Another study revealed that the inclusion of thyme, marjoram and sage in yoghurt had a stimulatory effect on the starter culture and the total viable counts and that the titratable acidity was increased in Labneh cheese samples fortified with essential oil. Moreover, thyme essential oils provided a stimulatory effect on the growth of the starter culture of *Lactobacillus casei* at a dose of 0.1 mL/100 g in yoghurt preparation [86]. However, in our study we noticed that dietary incorporation of cornus extract alone or with oregano and thyme essential oils provided a significant inhibitory effect on *Streptococcus thermophilus* populations.

The observed changes in feed utilization and FA concentrations could partly be due to the antioxidative effects of the plant extracts used or to changes in rumen microbial populations. A causal explanation of how EO mixtures affect ruminal fermentation patterns through microorganism modification has not been clearly established. Evans and Martin [87] examined the effects of thymol on ruminal microorganisms. Thymol, which is a constituent of the EO mixture tested here, inhibited gram-positive bacteria (*Streptococcus bovis* at 180 µg of thymol/mL and gram-negative bacteria (*Selenomonas ruminantium*)) at 90 µg of thymol/mL. Although these in vitro concentrations are much greater than the theoretically expected concentration of the mixtures tested in our experiment, bacterial inhibition could be a potential cause of the altered values of the fatty acid profiles of milk, yoghurt and cheese observed in the current study, which may be due to differences in fermentation patterns and subsequent biohydrogenation of the unsaturated fatty acids by rumen bacteria.

5. Conclusions

A clear finding of the current study is that the composition of herbal mixtures can greatly influence the performance outcomes of dairy sheep reared under stress conditions. Heat stress is a major issue in Mediterranean sheep farming as it may negatively affect milk yield and composition. Although the Chios breed and its crossbreeds are maintained and adapted to temperate climatic conditions, they also suffer when environmental temperature exceeds the thermoneutral zone. The current study suggests that dietary antioxidant compounds included in cornus, oregano and thyme have a positive impact on heat stress reduction and redox homeostasis. The addition of a specific cornus mixture with EO compounds of oregano and thyme had beneficial effects on milk production, urea concentration and SCC in milk samples of dairy ewes of the Chios crossbreed in the hot summer period. The addition of cornus hydrodistillation extract, as well as oregano and thyme essential oils, improved the oxidative stability of milk, cheese and yoghurt. However, the fatty acid profile in milk, cheese and yoghurt was not affected by the dietary addition of cornus extract and oregano and thyme EOs during the third month of lactation. Future research efforts should focus on defining specific types and compositions of various plant extracts and EOs in order to optimize their level of addition, thus resulting in favorable and consistent responses and effectively utilizing their benefits.

Author Contributions: Conceptualization, I.G. and I.M.; methodology, E.S., O.T., I.M., TM., Z.B., S.V. and D.L.; software, S.V., I.M., Z.B. and D.L.; validation, I.M., T.M. and D.L.; investigation, K.K. and S.C.; resources, S.C.; data curation, I.M., O.T. and D.L.; writing—original draft preparation, I.G. and I.M., K.K. conducted the experiment and collected samples. All authors have read, reviewed and agreed to the published version of the manuscript. All authors have read and agreed to the published version of the manuscript.

Funding: This research was funded by ORIZON Diatrofiki, grant number 99864, Research Committee Aristotle University.

Institutional Review Board Statement: This trial was carried out according to the regulations of the Greek Public Veterinary Service and approved by the Research Committee of Aristotle University of Thessaloniki, under the project numbered 99864.

Data Availability Statement: The data presented in this study are available on request from the corresponding author. The data are not publicly available due to privacy of Orizon Diatrofiki.

Conflicts of Interest: S.C. provided the plant material and had no role in the design of the study; in the collection, analyses, or interpretation of data. All other authors declare no conflict of interest.

References

1. Zervas, G.; Tsiplakou, E. The effect of feeding systems on the characteristics of products from small ruminants. *Small Rumin. Res.* **2011**, *101*, 140–149. [CrossRef]
2. Albright, J.L.; Tuckey, S.L.; Woods, G.T. Antibiotics in milk—A review. *J. Dairy Sci.* **1961**, *44*, 779–807. [CrossRef]
3. Barwary, M.S.Q.; Merkhan, K.Y.; Buti, E.T.S.; Isa, R.H.; Mustafa, K.N.; Yatem, C.A. Evaluation of medicinal plants (*Astragalus eriocephalus* and *Quercus infectoria*) as feed additives in awassi ewes' ration. *Iraqi J. Agric. Sci.* **2019**, *50*, 526–533.
4. Benchaar, C.; Calsamiglia, S.; Chaves, A.V.; Fraser, G.R.; Colombatto, D.; McAllister, A.; Beauchemin, K.A. A review of plant-derived essential oils in ruminant nutrition and production. *Anim. Feed Sci. Technol.* **2008**, *145*, 209–228. [CrossRef]
5. Caroprese, M.; Ciliberti, M.; Albenzio, M. Application of aromatic plants and their extracts in dairy animals. In *Feed Additives: Aromatic Plants and Herbs in Animal Nutrition and Health*; Florou-Paneri, P., Christaki, E., Giannenas, I., Eds.; Elsevier: London, UK, 2019; pp. 239–255.
6. Hendawy, A.O.; Mansour, M.M.; El-Din, A.M.N. Effects of medicinal plants on haematological indices, colostrum, and milk composition of ewes. *J. Vet. Med. Anim. Sci.* **2019**, *2*, 1–5.
7. Christaki, E.; Giannenas, I.; Bonos, E.; Florou-Paneri, P. Innovative uses of aromatic plants as natural supplements in nutrition. In *Feed Additives: Aromatic Plants and Herbs in Animal Nutrition and Health*; Florou-Paneri, P., Christaki, E., Giannenas, I., Eds.; Elsevier: London, UK, 2019; pp. 19–31.
8. Sallama, S.M.A.; Abdelgaleilb, S.A.M.; Buenoc, I.C.S.; Nassera, M.E.A.; Araujod, R.C.; Abdallac, A.L. Effect of some essential oils on in vitro methane emission. *Arch. Anim. Nutr.* **2011**, *65*, 203–214. [CrossRef] [PubMed]
9. Simitzis, P.E. Enrichment of animal diets with essential oils—A great perspective on improving animal performance and quality characteristics of the derived products. *Medicines* **2017**, *4*, 35. [CrossRef]
10. Dean, M.; Lampila, P.; Shepherd, R.; Arvola, A.; Saba, A.; Vassallo, M.; Claupein, E.; Winkelmann, M.; Lähteenmäki, L. Perceived relevance and foods with health-related claims. *Food Qual. Prefer.* **2012**, *24*, 129–135. [CrossRef]
11. Giannenas, I.; Sidiropoulou, E.; Bonos, E.; Christaki, E.; Florou-Paneri, P. The history of herbs, medicinal and aromatic plants, and their extracts: Past, current situation and future perspectives. In *Feed Additives: Aromatic Plants and Herbs in Animal Nutrition and Health*; Florou-Paneri, P., Christaki, E., Giannenas, I., Eds.; Elsevier: London, UK, 2019; pp. 1–15.
12. Anifantakis, E.M. *Feta and Related Cheeses*; Woodhead Publishing: London, UK, 1996.
13. Sarantinopoulos, P.; Kalantzopoulos, G.; Tsakalidou, E. Effect of *Enterococcus faecium* on microbiological, physicochemical and sensory characteristics of Greek feta cheese. *J. Food Microbiol.* **2002**, *76*, 93–105. [CrossRef]
14. Anifantakis, E.M.; Moatsou, G. *Feta and Other Balkan Cheeses. Production Methods, Manufacturing Stages and Properties*; Oxford Blackwell Publishing Ltd.: Oxford, UK, 2006.
15. Paraskevakis, N. Effects of dietary dried Greek oregano (*Origanum vulgare ssp. hirtum*) supplementation on blood and milk enzymatic antioxidant indices, on milk total antioxidant capacity and on productivity in goats. *Anim. Feed Sci. Technol.* **2015**, *209*, 90–97. [CrossRef]
16. El-Ghousein, S.S. Effect of some medicinal plants as feed additives on lactating awassi ewe performance, milk composition, lamb growth and relevant blood items. *J. Anim. Prod.* **2010**, *47*, 37–49.
17. Khamisabadi, H.; Fazaeli, H.; Ayasan, T. Effect of *Thymus vulgaris* or *Mentha peppermint* on lactating sanjabi ewe performance, milk composition, lamp growing and relevant blood metabolites. *J. Med. Plants ByProd.* **2020**, *10*, 95–101.
18. Wei, L.Y.; Gomaa, W.M.S.; Ametaj, B.N.; Alexander, T.W.; Yang, W.Z. Feeding red osier dogwood (*Cornus sericea*) to beef heifers fed a high-grain diet affected feed intake and total tract digestibility. *Anim. Feed Sci. Technol.* **2019**, *247*, 83–91. [CrossRef]
19. Gomaa, W.M.S.; Wei, L.Y.; Mosaad, G.M.; Aamer, H.; Alexander, T.W.; Yang, W.Z. In situ ruminal digestibility of red osier dogwood in finishing beef heifers. *Can. J. Anim. Sci.* **2018**, *98*, 888–892. [CrossRef]
20. Diamanti, A.; Igoumenidis, P.; Mourtzinos, I.; Yannakopoulou, K.; Karathanos, V. Green extraction of polyphenols from whole pomegranate fruit using cyclodextrins. *Food Chem.* **2017**, *214*, 61–66. [CrossRef] [PubMed]
21. Loukri, A.; Tsitlakidou, P.; Goula, A.; Assimopoulou, A.N.; Kontogiannopoulos, K.N.; Mourtzinos, I. Green extracts from coffee pulp and their application in the development of innovative brews. *Appl. Sci.* **2020**, *10*, 6982. [CrossRef]
22. McDowell, R.E. *Improvement of Livestock Production in Warm Climates*; Freeman: San Francisco, CA, USA, 1972.
23. Finch, V.A. *Heat as a Stress Factor in Herbivores under Tropical Conditions*; The Science Press: Graighall, South Africa, 1984.
24. Hayes, B.J.; Carrick, M.; Bowman, P.; Goddard, M.E. Genotype × environment interaction for milk production of daughters of Australian dairy sires from test-day records. *J. Dairy Sci.* **2003**, *86*, 3736–3744. [CrossRef]

25. Finocchiaro, R.; Kaam, J.B.C.H.M.; Portolano, B.; Misztal, I. Effect of heat stress on production of Mediterranean dairy sheep. *J. Dairy Sci.* **2005**, *88*, 1855–1864. [CrossRef]
26. Halliwell, B.; Whiteman, M. Measuring reactive species and oxidative damage in vivo and in cell culture: How should you do it and what do the results mean? *Br. J. Pharmacol.* **2004**, *142*, 231–255. [CrossRef]
27. Gladine, C.; Rock, E.; Morand, C.; Bauchart, D.; Durand, D. Bioavailability and antioxidant capacity of plant extracts rich in polyphenols, given as a single acute dose, in sheep made highly susceptible to lipoperoxidation. *Br. J. Nutr.* **2007**, *98*, 691–701. [CrossRef]
28. AOAC. *Official Methods of Analysis*, 15th ed.; AOAC: Arlington, VA, USA, 1990.
29. INRA. *Feeding System for Ruminants*; Wageningen Academic Publishers: Wageningen, The Netherlands, 2018.
30. Arnous, A.; Makris, D.P.; Kefalas, P. Correlation of pigment and flavanol content with antioxidant properties in selected aged regional wines from Greece. *J. Food Compos. Anal.* **2002**, *15*, 655–665. [CrossRef]
31. Giusti, M. Characterization of anthocyanin-rich waste from purple corncobs (*Zea mays* L.) and its application to color milk. *J. Agric. Food Chem.* **2005**, *53*, 8775–8781.
32. Wrolstad, P.; Dursta, R.; Lee, J. Tracking color and pigment changes in anthocyanin products. *Trends Food Sci. Technol.* **2005**, *16*, 423–428. [CrossRef]
33. Massada, Y. *Analysis of Essential Oils by Gas Chromatography and Spectrometry*; Wiley & Sons: New York, NY, USA, 1976.
34. Adams, R. *Identification of Essential Oil Components by Gas Chromatography/Mass Spectroscopy*, 4th ed.; Allured Publishing: Carol Stream, IL, USA, 2007.
35. Singleton, V.L.; Orthofer, R.; Lamuela-Raventios, R.M. Analysis of total phenols and other oxidation substrates and antioxidants by means of Folin-Ciocalteu reagent. *Methods Enzymol.* **1999**, *299*, 152–178.
36. Peperidou, A.; Kapoukranidou, D.; Kontogiorgis, C.; Hadjipavlou-Litina, D. Multitarget molecular hybrids of cinnamic acids. *Molecules* **2014**, *19*, 20197–20226. [CrossRef]
37. IDF. *Determination of Dry Matter in Cheese and Processed Cheese*; Standard 4; International Dairy Federation: Brussels, Belgium, 1958.
38. FIL-IDF. *Cheese-Determination of Chloride Content (Reference Method)*; FIL-IDF: Brussels, Belgium, 1972.
39. ISO. *Cheese. Determination of Fat Content. Gerber van Gulik Method*; Standard 3433; International Standards Organisation: Geneva, Switzerland, 1975.
40. Bligh, E.G.; Dyer, W.J. A rapid method of total lipid extraction and purification. Can. *J. Biochem. Physiol.* **1959**, *37*, 911–917.
41. International Organization for Standardization (ISO). *Milk Fat. Preparation of Fatty Acid Methyl Esters*; Standard 15884; ISO: Geneva, Switzerland, 2002.
42. Papaloukas, L.; Sinapis, E.; Arsenos, G.; Kyriakou, G.; Basdagianni, Z. Effect of season on fatty acid and terpene profiles of milk from Greek sheep raised under a semi-extensive production system. *J. Dairy Res.* **2016**, *83*, 375–382. [CrossRef] [PubMed]
43. Ulbricht, T.L.V.; Southgate, D.A.T. Coronary heart disease: Seven dietary factors. *Lancet* **1991**, *338*, 985–992. [CrossRef]
44. Patsoukis, N.; Zervoudakis, G.; Panagopoulos, N.T.; Georgiou, C.D.; Angelatou, F.; Matsokis, N.A. Thiol redox state (TRS) and oxidative stress in the mouse hippocampus after pentylenetetrazol-induced epileptic seizure. *Neurosci. Lett.* **2004**, *357*, 83–86. [CrossRef]
45. Wallace, R.J. Antimicrobial properties of plant secondary metabolites. *Proc. Nutr. Soc.* **2004**, *63*, 621–629. [CrossRef] [PubMed]
46. Wallace, R.J.; Colombatto, D.; Robinson, P.H. Enzymes, direct-fed microbials and plant extracts in ruminant nutrition. *Anim. Feed Sci. Tech.* **2008**, *145*, 1–4. [CrossRef]
47. Caroprese, M.; Ciliberti, M.G.; Albenzio, M.; Marino, R.; Santillo, A.; Sevi, A. Role of antioxidant molecules in milk of sheep. *Small Rum. Res.* **2019**, *170*, 102–108. [CrossRef]
48. Ribeiro, A.D.B.; Ferraz Junior, M.V.C.; Polizel, D.M.; Miszura, A.A.; Gobato, L.G.M.; Barroso, J.P.R.; Susin, I.; Pires, A.V. Thyme essential oil for sheep: Effect on rumen fermentation, nutrient digestibility, nitrogen metabolism, and growth. *Arq. Bras. Med. Vet. Zootec.* **2019**, *71*, 2065–2074. [CrossRef]
49. Zhou, R.; Wu, J.; Zhang, L.; Liu, L.; Casper, D.P.; Jiao, T.; Liu, T.; Wang, J.; Lang, X.; Song, S.; et al. Effects of oregano essential oil on the ruminal pH and microbial population of sheep. *PLoS ONE* **2019**, *14*, e0217054. [CrossRef]
50. Cappai, M.G.; Aboling, S. Toxic or harmful components of aromatic plants in animal nutrtion. In *Feed Additives: Aromatic Plants and Herbs in Animal Nutrition and Health*; Florou-Paneri, P., Christaki, E., Giannenas, I., Eds.; Elsevier: London, UK, 2019; pp. 147–158.
51. Akers, R.M. *Lactation and the Mammary Gland*; Iowa State Press: Ames, IA, USA, 2016.
52. El-Hawy, A.S. Biochemical, hematological, immunological responses and growth performance of Barki lambs born to ewes fed on *Nigella sativa* meal. *Res. J. Anim. Vet. Sci.* **2018**, *10*, 25–36.
53. Galbat, S.A.; El-Shemy, A.M.; Madpoli, M.A.; Omayma, E.L.; Maghraby, E.; Mossalami, L. Effects of some medicinal plants mixture on milk performance and blood components of Egyptian dairy goats. *Middle East J. Appl. Sci.* **2014**, *4*, 942–948.
54. Nieto, G. A review on applications and uses of thymus in the food industry. *Plants* **2020**, *9*, 961. [CrossRef] [PubMed]
55. Teixeira, B.; Marques, A.; Ramos, C.; Serrano, C.; Matos, O.; Neng, N.R.; Nogueira, G.M.F.; Saraiva, J.A.; Nunes, M.L. Chemical composition and bioactivity of different oregano (*Origanum vulgare*) extracts and essential oil. *J. Sci. Food Agric.* **2013**, *93*, 2707–2714. [CrossRef] [PubMed]
56. Nair, G.G.; Nair, C.K.K. Radioprotective effects of gallic acid in mice. *BioMed Res. Int.* **2013**, *2013*, 1–13. [CrossRef] [PubMed]

57. Santos, M.; Robinson, B.; Williams, P.H.; Losa, R. Effects of addition of an essential oil complex to the diet of lactating dairy cows on whole tract digestion of nutrients and productive performance. *Anim. Feed Sci. Technol.* **2010**, *157*, 64–71. [CrossRef]
58. Benchaar, C. Feeding oregano oil and its main component carvacrol does not affect ruminal fermentation, nutrient utilization, methane emissions, milk production, or milk fatty acid composition of dairy cows. *J. Dairy Sci.* **2020**, *103*, 1516–1527. [CrossRef]
59. Simitzis, P.; Massouras, T.; Goliomytis, M.; Charismiadou, M.; Moschou, K.; Economou, C.; Papadedes, V.; Lepesiotia, S.; Deligeorgis, S. The effects of hesperidin or naringin dietary supplementation on the milk properties of dairy ewes. *J. Sci. Food Agric.* **2019**, *99*, 6515–6521. [CrossRef]
60. Gessner, D.K.; Koch, C.; Romberg, F.J.; Winkler, A.; Dusel, G.; Herzog, E. The effect of grape seed and grape marc meal extract on milk performance and the expression of genes of endoplasmic reticulum stress and inflammation in the liver of dairy cows in early lactation. *J. Dairy Sci.* **2015**, *98*, 8856–8868. [CrossRef] [PubMed]
61. Winkler, A.; Gessner, D.K.; Koch, C.; Romberg, F.J.; Dusel, G.; Herzog, E. Effects of a plant product consisting of green tea and curcuma extract on milk production and the expression of hepatic genes involved in endoplasmic stress response and inflammation in dairy cows. *Arch. Anim. Nutr.* **2015**, *69*, 425–441. [CrossRef]
62. Isaak, C.K.; Petkau, J.K.; Karmin, O.; Ominski, K.; Rodriguez-Lecompte, J.C.; Siow, Y.L. Seasonal variations in phenolic compounds and antioxidant capacity of *Cornus stolonifera* plant material: Applications in agriculture. *Can. J. Plant Sci.* **2013**, *93*, 725–734. [CrossRef]
63. Makkar, H.P.S.; Siddhuraju, P.; Becker, K. Glucosinolates. In *Plant Secondary Metabolites. Methods in Molecular Biology*; Makkar, H.P.S., Siddhuraju, P., Becker, K., Eds.; Humana Press Inc.: Totowa, NJ, USA, 2007; pp. 55–60.
64. BenSaad, L.A.; Kim, K.H.; Quah, C.C.; Kim, W.R.; Shahimi, M. Anti-Inflammatory potential of ellagic acid, gallic acid and punicalagin A&B isolate from *Punica granatum*. *BMC Complement. Altern. Med.* **2017**, *17*, 47.
65. González-Segovia, R.; Quintanar, Q.L.; Salinas, E.; Ceballos-Salazar, R.; Aviles-Jiménez, F.; Torres-López, J. Effect of the flavonoid quercetin on inflammation and lipid peroxidation induced by *Helicobacter pylori* in gastric mucosa of guinea pig. *J. Gastroenterol.* **2008**, *43*, 441–447. [CrossRef] [PubMed]
66. Moldovan, B.; Popa, A.; David, L. Effects of storage temperature on the total phenolic content of cornelian cherry (*Cornus mas* L.) fruits extracts. *J. Appl. Bot. Food Qual.* **2016**, *89*, 208–211.
67. Scales, R. Anti-Oxidant Properties of Cornus Sericea. U.S. Patent US20150093460A1, 5 May 2020.
68. Braun, H.S.; Schrapers, K.T.; Mahlkow-Nerge, K.; Stumpff, F.; Rosendah, J. Dietary supplementation of essential oils in dairy cows: Evidence for stimulatory effects on nutrient absorption. *Animal* **2018**, *13*, 518–523. [CrossRef]
69. Santos, J.E.P.; Lean, I.J.; Golder, H.; Block, E. Meta-Analysis of the effects of prepartum dietary cation-anion difference on performance and health of dairy cows. *J. Dairy Sci.* **2019**, *102*, 2134–2154. [CrossRef] [PubMed]
70. Benchaar, C.; Petit, H.V.; Berthiaume, R.; Whyte, T.D.; Chouinard, P.Y. Effects of dietary addition of essential oils and monensin premix on digestion, ruminal fermentation characteristics, milk production, and milk composition in dairy cows. *J. Dairy Sci.* **2006**, *89*, 4352–4364. [CrossRef]
71. Tsiplakou, E.; Mountzouris, K.C.; Zervas, G. The effect of breed, stage of lactation and parity on sheep milk fat CLA content under the same feeding practices. *Livest. Sci.* **2006**, *105*, 162–167. [CrossRef]
72. Boutoial, K.; García, V.; Rovira, S.; Ferrandini, E.; Abdelkhalek, O.; López, M.B. Effect of feeding goats with distilled and non-distilled thyme leaves (*Thymus zygis* subp. *gracilis*) on milk and cheese properties. *J. Dairy Res.* **2013**, *80*, 448. [CrossRef]
73. Ludwiczuk, A.; Skalicka-Woniak, K.; Georgiev, M.I. Terpenoids. In *Pharmacognosy, Badal, S., Delgoda, R., Eds.*; Academic Press: Cambridge, MA, USA, 2017; pp. 233–266.
74. Liu, L.L.; He, J.H.; Xie, H.B.; Yang, Y.S.; Li, J.C.; Zou, Y. Resveratrol induces antioxidant and heat shock protein mRNA expression in response to heat stress in black-boned chickens. *Poult. Sci.* **2014**, *93*, 54–62. [CrossRef]
75. Steinmetz, K.A.; Potter, J.D. Vegetables, fruit, and cancer prevention: A review. *J. Am. Diet. Assoc.* **1996**, *96*, 1027–1039. [CrossRef]
76. Knekt, P.; Kumpulainen, J.; Järvinen, R.; Rissanen, H.; Heliövaara, M.; Reunanen, A.; Hakulinen, T.; Aromaa, A. Flavonoid intake and risk of chronic diseases. *Am. J. Clin. Nutr.* **2002**, *76*, 560–568. [CrossRef]
77. Sarangi, S. Adaptability of goats to heat stress: A review. *Pharma Innov. J.* **2018**, *7*, 1114–1126.
78. Havlin, J.M.; Robinson, P.H. Intake, milk production and heat stress of dairy cows fed a citrus extract during summer heat. *Anim. Feed Sci. Technol.* **2015**, *208*, 23–32. [CrossRef]
79. Basdagianni, Z.; Banos, G.; Abas, Z.; Arsenos, G. Estimation of daily and total lactation milk yield of Chios ewes from single morning or evening records. *Livest. Prod. Sci.* **2005**, *92*, 59–68. [CrossRef]
80. Langeveld, W.T.; Veldhuizen, E.J.A.; Burt, S.A. Synergy between essential oil components and antibiotics: A review. *Crit. Rev. Microbiol.* **2014**, *40*, 76–94. [CrossRef]
81. Hassanpour, S.; Maheri-Sis, N.; Eshratkhah, B.; Mehmandar, F.B. Plants and secondary metabolites (Tannins): A Review. *Int. J. For. Soil Eros.* **2011**, *1*, 47–53.
82. Demir, H.U.; Atalay, D.; Erge, H.S. Kinetics of the changes in bio-active compounds, antioxidant capacity and color of Cornelian cherries dried at different temperatures. *J. Food Meas. Charact.* **2019**, *13*, 2032–2040. [CrossRef]
83. Pantelidis, G.E.; Vasilakakis, M.; Manganaris, G.A.; Diamantidis, G. Antioxidant capacity, phenol, anthocyanin and ascorbic acid contents in raspberries, blackberries, red currants, gooseberries and cornelian cherries. *Food Chem.* **2007**, *102*, 777–783. [CrossRef]
84. Barkat, M.; Laib, I. Antioxidant activity of the essential oil from the flowers of *Lavandula stoechas*. *J. Pharmacogn. Phytother.* **2012**, *4*, 96–101.

85. Ducková, V.; Kročko, M.; Kňazovická, V.; Čanigová, M. Evaluation of yoghurts with thyme, thyme essential oil and salt. *Acta Univ. Agric. Silvic. Mendelian. Brun.* **2018**, *66*, 365–369. [CrossRef]
86. Al Otaibi, M.; El Demerdash, H. Improvement of the quality and shelf life of concentrated yoghurt (labneh) by the addition of some essential oils. *Afr. J. Microbiol. Res.* **2008**, *2*, 156–161.
87. Evans, J.; Martin, S. Effects of thymol on ruminal microorganisms. *Curr. Microbiol.* **2000**, *41*, 336–340. [CrossRef]

animals MDPI

Article

The Effects of Replacing Soybean Meal with Rapeseed Meal, Cottonseed Cake, and Fava Beans on the Milk Yield and Quality Traits in Milking Ewes

Aphrodite I. Kalogianni [1], Marios Moschovas [1], Foteini Chrysanthakopoulou [2], Thomai Lazou [3], Georgios Theodorou [4], Ioannis Politis [4], Ioannis Bossis [5] and Athanasios I. Gelasakis [1,*]

1 Laboratory of Anatomy and Physiology of Farm Animals, Department of Animal Science, School of Animal Biosciences, Agricultural University of Athens (AUA), Iera Odos 75 Str., 11855 Athens, Greece; afrokalo@aua.gr (A.I.K.); moschovas@aua.gr (M.M.)
2 Agricultural Cooperative 'Agrinio Union', Papaioannou 24 Str., 30131 Agrinio, Greece; chrysanthakopoulouf@gmail.com
3 Laboratory of Hygiene of Foods of Animal Origin—Veterinary Public Health, School of Veterinary Medicine, Faculty of Health Sciences, Aristotle University of Thessaloniki, 54124 Thessaloniki, Greece; tlazou@vet.auth.gr
4 Laboratory of Animal Husbandry, Department of Animal Science, School of Animal Biosciences, Agricultural University of Athens (AUA), Iera Odos 75 Str., 11855 Athens, Greece; gtheod@aua.gr (G.T.); i.politis@aua.gr (I.P.)
5 Laboratory of Animal Husbandry, Department of Agricultural Sciences, School of Agriculture, Forestry and Natural Resources, Aristotle University of Thessaloniki (AUTH), 54124 Thessaloniki, Greece; bossisi@agro.auth.gr
* Correspondence: gelasakis@aua.gr

Citation: Kalogianni, A.I.; Moschovas, M.; Chrysanthakopoulou, F.; Lazou, T.; Theodorou, G.; Politis, I.; Bossis, I.; Gelasakis, A.I. The Effects of Replacing Soybean Meal with Rapeseed Meal, Cottonseed Cake, and Fava Beans on the Milk Yield and Quality Traits in Milking Ewes. *Animals* **2022**, *12*, 274. https://doi.org/10.3390/ani12030274

Academic Editors: Monica Isabella Cutrignelli and Jacek Wójtowski

Received: 28 November 2021
Accepted: 20 January 2022
Published: 22 January 2022

Simple Summary: The substitution of soybean meal in farm animal diets is considered vital for the economic and environmental sustainability of the livestock sector. However, data regarding the effects of a soybean meal replacement on the milk yield and quality traits in dairy sheep are scarce. In our study, two isonitrogenous and isoenergetic diets were used, with soybean meal of a typical ration being replaced by a mixture of rapeseed meal, cottonseed cake, and fava beans. The milk yield and the body condition scores were recorded, and milk samples were analyzed monthly for their fat, protein, lactose, and total solids yields, as well as for somatic cell counts, total bacterial counts, pH, electrical conductivity, and the refractive index. Daily and 100-day fat yields were significantly increased in the group fed the experimental ration and the electrical conductivity was significantly decreased in the same group, while no adverse effects on any of the rest of the studied milk production traits were observed.

Abstract: The replacement of soybean meal (SBM) from intensively reared dairy sheep diets has emerged as a significant challenge for sustainable production. However, the effects of this replacement on milk production have not been sufficiently elucidated. The objective of this study was to prospectively assess the effects of replacing SBM with a mixture of alternative protein sources on the milk yield (MY) and the milk quality traits (MQT) in intensively reared dairy sheep. A total of 112 multiparous, purebred milking ewes of the Chios and Frizarta breeds, from two intensive dairy sheep farms, were involved in the study, postweaning, and were assigned to either the control (CR) or the experimental ration (ER) group. In the ER, 3/4 of the SBM was replaced by a mixture of rapeseed meal, cottonseed cake, and fava beans, producing a ration of a similar nutritional value. MY, MQT, and body condition scores were recorded for each individual ewe monthly for a period of 4 months during lactation. The experimental ration was associated with beneficial effects on daily and 100-day fat yields and on the electrical conductivity of milk as an improved udder health status indicator, with no adverse effects on any of the rest of the studied milk production traits.

Keywords: soybean meal; fava beans; rapeseed meal; cottonseed cake; dairy sheep; milk yield; milk quality

1. Introduction

The demand for sheep milk and products thereof (e.g., cheese, yoghurt, and butter) has increased over the years due to their perceived high nutritional value and the consumer demands to produce niche and premium-quality dairy products [1]. This demand-driven evolution of the sheep milk processing sector has dragged the tendency towards the intensification of production and the modernization of husbandry systems, mainly in the developed world, as exemplified by European countries in the Mediterranean basin (i.e., Greece, Italy, France, and Spain). Among the factors affecting the sustainability of these systems, evidence-based nutrition and precision feeding remain the cornerstones supporting the sufficient exploitation of highly productive dairy sheep breeds [2] in regard to their milk (i) quantity and quality traits (e.g., milk yield, protein, fat, lactose, total solids, fatty acid profile, etc.), (ii) technological and coagulation properties, and (iii) organoleptic traits.

In general, dairy ewe nutrition is characterized by increased demands in energy and protein during the milking period. Particularly in intensive farms, nutrients are supplied inside the barn by feeding concentrates of high nutritional value, gradually transforming grazing-oriented traditional sheep farming systems to zero-grazing indoor systems [3]. Soybean meal (SBM) currently constitutes the most widely used protein-rich feedstuff in the livestock sector for meat and milk production. It is the co-product of soybean oil extraction and represents approximately 70% of the consumed oilseed meals globally [4]. It is highly preferred in diets of dairy ruminants due to its high crude protein (CP) content (44–56% of dry matter (DM)) and nitrogen digestibility (about 80%); it also contains crude fiber (CF) ca. 1.5–6.0% of DM, fat ca. 2% of DM, and 2.0 Mcal/kg net energy for lactation (NEL) [5]. Despite the unquestionable feeding value of SBM, its partial or total substitution in farm animal diets has emerged as an imperative need due to logistic, economic, and environmental burdens. The USA, Brazil, and Argentina rank first in the list of the SBM producing and exporting countries, continuously intensifying their production despite the recognized environmental impacts imposed by its cultivation [6]. Interestingly, more than 40% of the global available SBM is exported to the EU due to the negligible self-sufficiency of the latter, via an economically and environmentally detrimental transatlantic trading system [7]. In addition, soybean is the most widely used genetically modified crop, opposing the consumer awareness of genetically modified organisms [8].

A combination of grain legumes and the by-products of oil plants are a promising alternative protein source in ruminant nutrition due to (i) their high nutritional value, (ii) the improvement of soil fertility and the reduction of nitrogenous fertilization induced by legume cultivation, (iii) their potential cultivation in less fertile, non-irrigated fields, (iv) the exploitation of industrial by-products within the circular agricultural economy model, and (v) the lack of competition with human nutrition [7,9]. The grain legumes of *Leguminosae* family (e.g., fava bean, pea, lupin) and oil plants (e.g., rape and cotton) have been studied as alternative feed resources in both monogastric and ruminant farm animals, though they display contradictory effects on their productivity [10–18]. In Greece, three popular, locally produced feedstuffs integrated into farm animal diets as protein sources are rapeseed meal, cottonseed cake, and fava beans. Rapeseed (*Brassica* spp.) constitutes a relatively new cultivation which has emerged mainly during the last decade in the country and has been exploited for the production of biodiesel and for its soil fertilizing capacity. Rapeseed meal is extensively used with meat and wool sheep as an efficient alternative to SBM, offering similar energy, digestibility, and protein degradability comparable to SBM [19]. Rapeseed meal is rich in protein (CP ca. 33–45% of DM) and fiber (CF ca. 9–18% of DM) content, and contains ca. 2% fat and 1.70 Mcal/kg NEL [20]. Cotton (*Gossypium* spp.) is a customary and extensively cultivated crop with a long tradition and experience in its cultivation in many parts of the country. Cottonseed cake, a by-product of the textile industry, is a valuable feedstuff for ruminants given its high protein content (CP ca. 20–50% of DM) and its resistance to gossypol toxicity, in contrast to monogastric animals [21]. Cottonseed cake also contains CF ca. 7–17% of DM, fat ca. 2–10%, and ca 2.0 Mcal/kg NEL [16,22]. Fava bean crop (*Vicia faba* L. minor) is suitable for cultivation in unfavored soils under less intensive and/or organic production systems, given its limited water and fertilizer demands and the consequent low environmental footprint. It is abundant in protein (CP ca. 25–35% of DM) and it contains CF ca. 9–11% of DM, fat ca. 2% of DM, and 1.70 Mcal/kg NEL [23,24]. Although its high ruminal nitrogen degradability and the presence of antinutritional factors (tannins and pyrimidine glycosides) have hindered its preference in intensive farming systems [25,26], the currently available low levels of tannins or tannin-free cultivars, and the implementation of technological treatments such as the extrusion, have improved its nutritional value and enhanced its potential use in ruminant diets [26,27].

Dairy sheep farming is the most dynamic livestock sector in Greece. In the last two decades, the sector has rapidly evolved to cover the increasing demands for the Protected Designation of Origin feta cheese and sheep yoghurt, with the intensification of farming systems being the driving force. Consequently, zero-grazing, high-input farms have emerged and have increased rapidly in the mainland. This transformation has led to the development of balanced diets and more sophisticated feeding protocols to efficiently meet the nutritional requirements of high-yielding animals within a reasonable production cost. Such diets include SBM as the main protein source, utilized to meet corresponding requirements of ewes during lactation, since the evidence to support its efficient substitution with alternative protein sources in terms of animal performance is lacking.

The hypothesis here was that partial replacement of SBM with other protein-rich grain legumes and by-products of oil plants has no adverse effects on the milk yield and quality. To test this hypothesis, the objective of this study was to prospectively evaluate the effects of the partial replacement of SBM with a mixture of rapeseed meal, cottonseed cake, and fava beans on the milk yield and milk quality traits (fat, protein, lactose, total solids yield, somatic cell count (SCC), total bacterial count (TBC), electrical conductivity (EC), refractive index (RI), and pH) in intensively reared dairy sheep of two indigenous Greek breeds.

2. Materials and Methods

2.1. Animals and Diets

Two intensive dairy sheep farms located at Aetolia-Acarnania in Western Greece were involved in the study. A total of 112 purebred, multiparous (2nd, 3rd and ≥4th parity), milking ewes at postweaning (50 days post-partum), namely, 64 Frizarta (Farm A) and 48 Chios (Farm B), were randomly selected and enrolled in a 4-month prospective study. Initially, in each farm, the selected ewes were homogeneously allocated into two equal groups (with 32 Frizarta and 24 Chios each, for Farm A and B, respectively) according to their parity number, their daily milk yield (DMY) and the milk quality traits (MQT, i.e., fat, protein, lactose, total solids yield, SCC, TBC, EC, RI, and pH) and were assigned to either the control (C) or the experimental (E) groups. The two groups were permanently housed in separate pens and were mechanically milked twice a day with a 12-h interval (8:00 a.m. and 8:00 p.m.), following the routine of the farm. In both groups, concentrates were fed in a pelleted form, whereas alfalfa hay (18% CP) and wheat straw of a similar nutritional value were equally supplied. The ration fed in the group C ewes was a typical one, incorporating 20% of SBM as the main protein source. In group E, 3/4 of SBM was replaced by a mixture of locally produced rapeseed meal, cottonseed cake, and fava beans to produce a ration of similar nutritional value. The assessment of the nutritional value of the two rations was conducted following standard procedures according to the Association of Official Agricultural Chemists (AOAC) in an accredited laboratory. The composition and the nutritional values of the two rations are presented in Table 1.

Table 1. Composition and chemical analysis of the rations fed in the control (C) and experimental (E) groups.

	Control Ration	Experimental Ration
Composition (%)		
Soybean meal	20.0	5.0
Rapeseed meal	-	13.0
Cottonseed cake	-	10.0
Fava beans	-	12.5
Barley grain	13.5	-
Corn grain	50.0	52.0
Wheat bran	14.0	-
Sugar beet pulp	-	5.0
Vitamins and minerals	2.5	2.5
Chemical analysis		
Dry matter (%)	86.73	88.18
Crude protein (% of DM)	15.63	15.96
Ash (% of DM)	4.32	4.91
Fat (% of DM)	3.19	3.51
NDF (% of DM)	8.00	13.45
ADF (% of DM)	1.76	6.74
ADL (% of DM)	0.04	1.86
Starch (% of DM)	50.63	39.82
Calcium (% of DM)	1.61	1.71
Phosphorus (% of DM)	0.52	0.45
Net energy for lactation † (Mcal per kg DM)	1.95	1.96

DM: dry matter; NDF: neutral detergent fiber; ADF: acid detergent fiber; ADL: acid detergent lignin; † theoretical estimation using the software Plurimix System® v.2.41.34, (Fabermatica, Ostiano, Italy).

During the study, the diets (concentrates and roughages) of the two groups were isocaloric and isonitrogenous and were adapted to meet the nutritional demands of the ewes according to their lactation stage and their milk production level. On the contrary, the fiber content (neutral detergent fiber (NDF), acid detergent fiber (ADF), and acid detergent lignin (ADL)) was lower in the group C ration (CR) compared to the group E ration (ER), while the starch content was higher in CR. The concentrate quantities ranged from 1.0 to 1.3 kg/ewe/day with alfalfa hay from 1.1 to 1.4 kg/ewe/day, regularly tuned to meet the nutrient requirements of the animals during the study. Quantities of concentrates were constantly adjusted to the milk yield of each ewe and the additional demanded dry matter for high yielding animals was individually provided in the milking parlor. In any case, in the two groups, the quantities of concentrates and alfalfa hay were equal for equally producing animals. Concentrates were provided in feeders, both during (in the milking parlor) and after milking (in the barn), twice a day, while roughages were fed twice a day after milking; all feed refusals were removed before the next feeding session.

2.2. Milk Sampling and Analyses

Following a 30-day adaptation period of the diets (at 80 days post-partum) the enrolled ewes in the two farms were prospectively studied monthly for 4 months, including the adaptation period. In every sampling occasion, the milk yield was recorded, and the milk samples were collected from each individual ewe and were transferred to the lab for chemical analyses. The milk yield recording and milk sampling were performed using ICAR (International Committee of Animal Recording)-approved equipment (Waikato Milkmeter, InterAg, Hamilton, New Zealand) and protocols during morning milking. Two milk samples (ca. 70 mL each) were collected per animal. The first was used for TBC estimation and was aseptically collected (proper udder and teat cleaning with antiseptic towels, discarding of first milk streams, and collection of composite samples from both half udders), before the milk yield recording. The second was collected from the milkmeter's sampler and was used for the rest of the analyses. Sodium azide (sodium azide tablets, Supelco®, Merck Milipore, Burlington, MA, USA) was added, and the samples were transferred under cool storage conditions (4 °C) and were analyzed within 24 h. Milk samples were analyzed for fat, protein, lactose, total solids contents (MilkoScan™ FT+, Foss, Hilleroed Denmark), SCC (Fossomatic™ FC, Foss, Hilleroed Denmark), and TBC (Bactoscan™ FC+, Foss, Hilleroed Denmark). Daily milk, fat, protein, lactose, and total solids yields were calculated using the morning milking records and were adjusted following the ICAR methods [28]. The physicochemical characteristics of milk samples, namely the pH, EC, and RI, were measured at 20 °C with a pen-type pHmeter-conductometer (EZDO 7200, GOnDO Electronic Co., LTD, Taipei, Taiwan) and a handheld refractometer (RHB-32ATC, Laxco, Inc., Mill Creek, WA, USA) according to the brix scale, respectively. At the end of the study, the total milk, fat, protein, lactose, and total solids yields were calculated for the 100 days of the experiment using the Fleischmann method and the ICAR recommendations [28]. Moreover, the body condition score (BCS) was recorded by the same veterinarian in each sampling using a five-degree scale (1 = emaciated, 5 = obese) [29].

2.3. Statistical Analyses

SPSS v23 software (IBM Corp., Armonk, NY, USA) was used for the statistical analyses, with the statistical significance being set at the 0.05 level. Initially, SCC and TBC were log-transformed, and the Kolmogorov–Smirnov test was used to test for normality. Descriptive statistics (mean ± standard error) were calculated for the milk quality and quantity traits for groups C and E throughout the study. The following mixed linear regression model was formulated for the assessment of the effects of the two diets on DMY and MQT:

$$Y_{ijklm} = \mu + F_i + G_j + P_k + S_l + a_1 \times BCS + E_m + \delta_{ml} + e_{ijklm} \text{ (model 1)}$$

where Y_{ijklm} = dependent variables (daily milk, fat, protein, lactose, total solids yield, logarithm of SCC (logSCC), logarithm of TBC (logTBC), EC, pH, and RI); μ = intercept; F_i = fixed effect of the farm (i = 2 levels; 0 = Farm A, 1 = Farm B); G_j = fixed effect of the ration (j = 2 levels; 0 = control ration, 1 = experimental ration); P_k = fixed effect of the parity number (k = 3 levels; 2nd, 3rd, ≥4th parity); S_l = fixed effect of the sampling occasion (l = 4 levels; 1st to 4th sampling occasion); a_1 = fixed effect of the regression coefficient of BCS; E_m = random variation of the mth ewe; δ_{ml} = repeated variation of the mth ewe in the lth sampling occasion; e_{ijklm} = residual error.

Akaike's information criterion (AIC) value was used for the selection of the most appropriate covariance structure in the mixed linear model and the first-order autoregressive was selected as the most appropriate one.

An analysis of covariance was used to assess the effects of the diet on the 100-day milk, fat, protein, lactose, and total solids yields, as described in the following model:

$$Y_{ijk} = \mu + F_i + G_j + P_k + e_{ijk} \text{ (model 2)}$$

where Y_{ijkl} = dependent variables (100-d milk, fat, protein, lactose and total solids yield); μ = intercept; F_i = fixed effect of the farm (i = 2 levels; 0 = Farm A, 1 = Farm B); G_j = fixed effect of the ration (j = 2 levels; 0 = control ration, 1 = experimental ration); P_k = fixed effect of the parity number (k = 3 levels; 2nd, 3rd, ≥4th parity); and e = residual error.

The assumptions of normal distribution, homoscedasticity, and linearity for the models were checked by the assessment of a scatterplot of standardized predicted values against the standardized residuals and the probability-probability and quantile-quantile plots of standardized residuals.

3. Results

3.1. Descriptive Statistics

Figure 1 demonstrates the progress of DMY and the studied MQT for the two groups during the study. The average DMY continuously decreased from the middle to the end of lactation in both groups, ranging from 1.4 to 0.9 l for group C ewes and from 1.3 to 0.9 l for group E ewes. A similar declining trend was observed for the daily fat, protein, lactose, and total solids yields. During the study, the mean daily fat, protein, lactose and total solids yields varied from 83.9 to 56.4 g, 75.2 to 51.1 g, 63.4 to 39.7 g, and 232.3 to 157.5 g in group C, respectively, while in group E the values varied from 93.2 to 58.5 g, 80.1 to 51.5 g, 63.0 to 39.7 g, and 249.1 to 157.5 g, respectively. Table 2 summarizes the mean values of daily and total milk yields and the milk quality traits. The mean values of BCS varied from 2.7 to 2.9 in both groups, following a similar trend during the study.

Table 2. Mean values (± SD) of daily and total milk yields and milk quality traits and the effects of diet on them (reference category for comparisons is group C).

	Dependent Variables	Group C Mean (± SE)	Group E Mean (± SE)	*B*	SEM	*p*-Value	95% CI Lower Bound	95% CI Upper Bound
Model (1)	Daily milk yield (L)	1.09 (0.28)	1.09 (0.28)	0.01	0.081	0.946	−0.15	0.14
	Daily fat yield (g)	67.64 (18.46)	76.20 (18.44)	8.55	3.945	0.032	0.73	16.38
	Daily protein yield (g)	62.30 (20.61)	66.11 (20.58)	3.81	4.219	0.369	−12.17	4.56
	Daily lactose yield (g)	51.30 (59.3)	52.44 (59.52)	1.14	3.686	0.757	−8.45	6.16
	Daily total solids yield (g)	189.64 (63.53)	205.04 (63.78)	15.40	12.020	0.203	−39.23	8.44
	Log of SCC (10^3/mL)	5.68 (0.45)	5.70 (0.45)	−0.02	0.092	0.824	−0.02	0.16
	Log of TBC (cfu × 10^3/mL)	4.50 (0.27)	4.49 (0.27)	−0.02	0.066	0.783	−0.11	0.15
	pH †	6.60 (0.09)	6.57 (0.09)	−0.04	0.021	0.068	0.00	0.08
	Electrical conductivity (mS/cm) †	3.58 (0.08)	3.39 (0.08)	−0.20	0.054	0.000	−0.31	−0.09
	Refractive index † (brix)	15.38 (0.57)	15.25 (0.61)	−0.12	0.685	0.858	−1.24	1.48
Model (2)	100-day milk yield (L)	109.57 (±5.45)	109.26 (±4.68)	−1.10	7.216	0.879	−15.41	13.21
	100-day fat yield (g)	6863.06 (±266.76)	7728.33 (±274.00)	799.96	378.095	0.037	50.35	1549.57
	100-day protein yield (g)	6161.33 (±297.23)	6508.12 (±275.18)	281.09	410.284	0.495	−532.34	1094.52
	100-day lactose yield (g)	5070.31 (±268.36)	5124.32 (±230.58)	4.47	355.361	0.990	−700.66	709.61
	100-day total solids yield (g)	18953.30 (±842.55)	20410.30 (±791.83)	1261.97	1161.835	0.280	−1041.48	3565.42

Group C: control group; Group E: experimental group; SE: standard error; B: coefficient; SEM: standard error of the mean of B coefficient; model (1): mixed linear regression model for daily milk yield and milk quality traits; model (2): linear regression model for 100-day milk yield and milk quality traits; SCC: somatic cell count; TBC: total bacterial count; cfu: colony-forming unit; † measured at 20 °C.

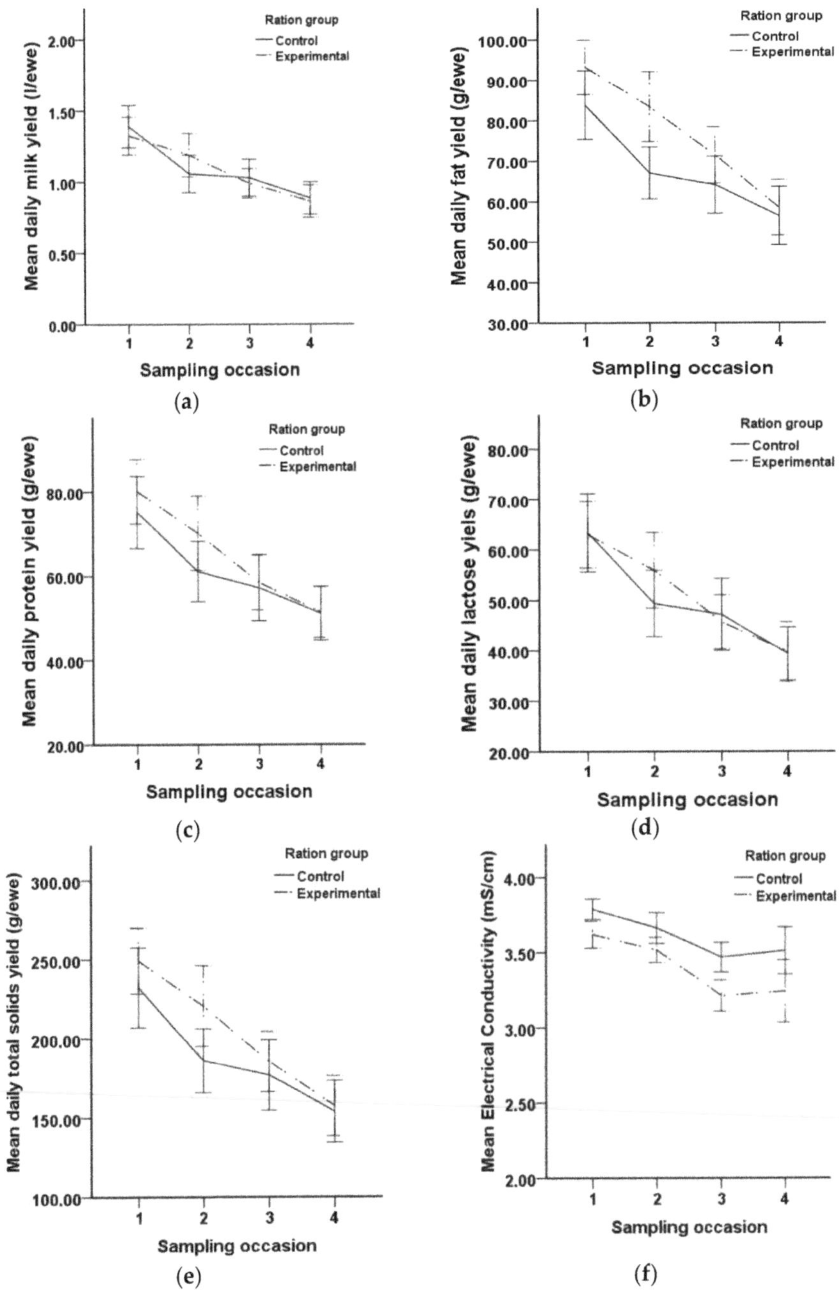

Figure 1. *Cont.*

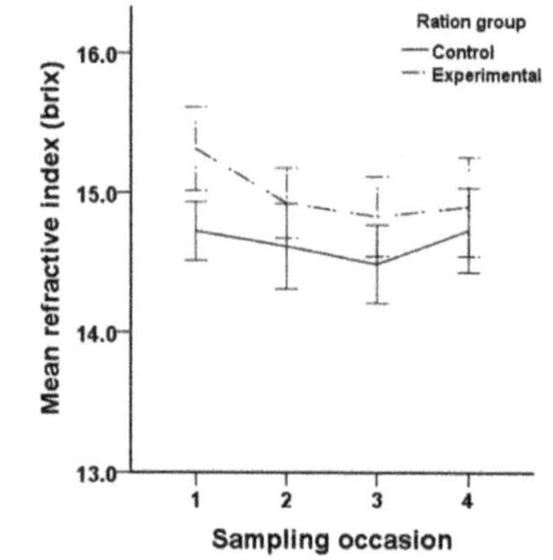

(g)

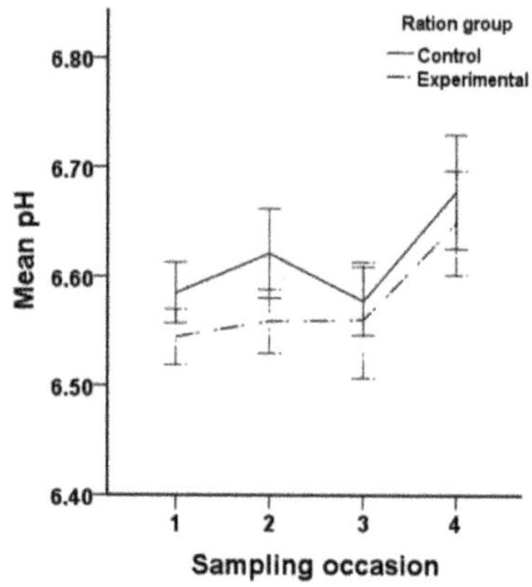

(h)

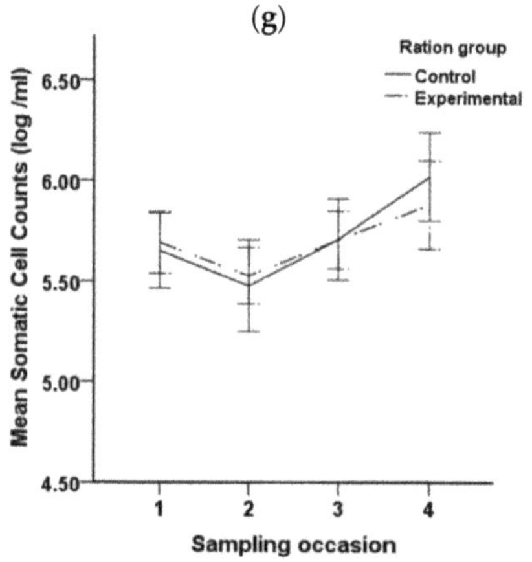

(i)

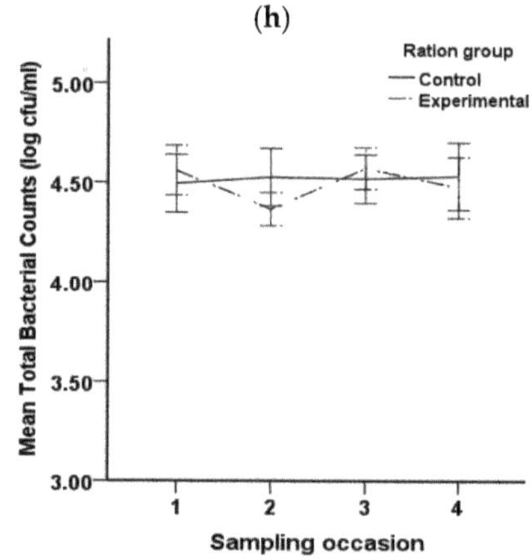

(j)

Figure 1. Mean values of (**a**) daily milk yield; (**b**) daily fat yield; (**c**) daily protein yield; (**d**) daily lactose yield; (**e**) daily total solids yield; (**f**) electrical conductivity; (**g**) refractive index; (**h**) pH; (**i**) logarithm of somatic cell counts; and (**j**) logarithm of total bacterial counts, for the two groups during the study.

3.2. *The Effects of Diet on the Daily and Total Milk Yields and Milk Quality Traits*

Table 2 summarizes the effects of diet on the daily and total yields of milk, fat, protein, lactose, total solids, logSCC, and logTBC, as well as on pH, EC, and RI. A significant effect was observed on the daily and total fat yields ($p = 0.032$ and $p = 0.037$, respectively) and on EC ($p < 0.001$). Namely, ewes in group E yielded more fat in their milk (daily ca. 8.6 g, 95% CI, 0.7 to 16.4 g, and total ca. 800 g, 95% CI 50.4 to 1549.57 g) compared to ewes in group C. Furthermore, milk EC was decreased by 0.2 mS/cm ($p < 0.001$, 95% CI, −0.3 to −0.1 mS/cm) in group E ewes.

3.3. *The Effects of Other Explanatory Variables on the Daily and Total Milk Yields and Milk Quality Traits*

The sampling occasion had a statistically significant effect ($p < 0.001$) in every case, except from the RI and the logarithm of TBC. Moreover, Farm A ewes had significantly higher logarithm of SCC and TBC ($p < 0.001$, increased by 0.44 and 0.30 logarithms, respectively) and an increased milk EC by 0.48 mS/cm ($p < 0.001$) compared to Farm B. The parity number had no statistically significant effect on any of the studied traits, while a one-degree increase on BCS was associated with a decrease in pH by 0.08 units ($p < 0.01$) and milk EC by 0.27 mS/cm. Regarding the total milk yield and the quality traits (fat, protein, lactose, and total solids yield), the farm and parity number had no statistically significant effects on any case, except from Farm B ewes that had a significantly higher total fat yield compared to Farm A ($p < 0.05$, increased by 1022.6 g).

4. Discussion

To the best of our knowledge, this is the first time the effects of SBM substitution with a mixture of rapeseed meal, cottonseed cake, and fava beans on milk performance in dairy sheep have been prospectively studied; no significant effects were observed on MY and MQT, with the exception of a favorable effect on the milk fat yield observed in the experimental ration. The combination of alternative protein sources aims towards the efficient coverage of the metabolizable protein requirements of dairy sheep through the improvement of the rumen degradable protein (RDP)-to-rumen undegradable protein (RUP) ratio and the balance of essential amino acids [15,30]. As the amino acid profile differs significantly among the protein sources, the ideal diet should be formulated by a panel of protein sources which complement microbial proteins with the essential amino acids for milk production, such as methionine, lysine, leucine, and histidine [31].

The majority of relevant studies in the available literature have assessed the effects of substituting SBM with a single alternative protein feed, mainly on the quantity and quality of milk in dairy cows. Therefore, extrapolating and directly comparing the results of these studies with our findings is not appropriate.

The components selected in the experimental ration, namely, rapeseed meal, cottonseed cake, and fava beans, are among the most commonly used alternative protein sources for the substitution of SBM in small ruminant diets in Greece, as the crops they derive from are popular in different regions around the country. However, the effectiveness of integrating them into the dairy sheep diets to effectively substitute SBM has not been assessed, until now, on an evidential basis and under commercial farming conditions. Up-to-date data demonstrating the effects of SBM replacement with rapeseed meal on MY and MQT in dairy sheep are scarce. On the contrary, relevant studies in dairy cows have extensively documented these effects and rapeseed meal was found to be more effective than SBM and other oilseed feeds, favoring their milk, protein, lactose, and fat yields without affecting BCS [31–35]. The decrease in milk urea and urine urea nitrogen, and the increase of the essential amino acid availability, such as histidine, methionine, and lysine, in cows fed rapeseed meal, restates SBM superiority in terms of ruminal degradability, protein digestibility, and nitrogen efficiency [32,34–36].

Cottonseed cakes have been exploited as an alternative to the SBM protein source mainly in monogastric farm animals, for fattening lambs, and for growing goats, and less commonly in adult ruminants. Therefore, data with which to compare the results of our study are limited. The addition of cottonseed cake in ostriches and broilers improved their growth rate [10,22], whereas in pigs, the lysine deficiency and the presence of gossypol adversely affected their performance [37]. The substitution of SBM with cottonseed cake in goat kids and lambs did not influence their growth or their feeding efficiency, as well as the microbial protein synthesis [16,38].

The substitution of SBM with fava beans has been studied in dairy ewes in Italy [39,40]. Liponi et al. [39] replaced soybean meal with fava beans or peas in 18 postweaning Massese-bred lactating ewes for 70 days, while Bonanno et al. [40] replaced maize grain and SBM with fava beans, chickpeas, or peas mixed with barley in 12 Comissana-bred lactating ewes for 21 days. In both cases, no statistically significant differences regarding MY, MQT, BCS, SCC, TBC, and pH were observed. On the contrary, in goats, fava beans improved the milk protein yield when compared to other feeds with a high protein content (e.g., sunflower meal, vetch, and bitter vetch) [41,42]. Relevant studies in dairy cows indicated that the partial [43] or total replacement of SBM with fava beans [44] or with a combination of fava beans and rapeseed meal [26] did not adversely affect the milk yield and composition, which is in accordance with our findings. However, the combination of fava beans with peas reduced the dry matter intake and milk yield in cows [12].

In the current study, daily and total milk fat yields were significantly increased (by ca. 12.0%) in the group E, compared to the group C, ewes. An obvious explanation could be the increased content of NDF against the starch content in the experimental ration, as previously reported in dairy sheep and cows [45–48]. Although the mechanisms and regulators of the milk fat synthesis are quite complicated and insufficiently evidenced in dairy sheep, high starch and low NDF content have been related to the disruption of the acetate-to-propionate and acetate-to-butyrate ratios, with a further effect on rumen fermentation and the production of fatty acids for milk fat synthesis [49,50]. In any case, other factors affecting the efficiency of the experimental rations need to be further elucidated and more studies under different experimental protocols are warranted to reveal potential differences on the ruminal metabolism and feed digestibility of the two rations, and to evidence the superiority claims of the experimental rations.

In our study, the mean values of BCS and their progress were similar in the two groups and followed the expected pattern for the studied period (mid to late lactation) [51]. The two diets succeeded in meeting the energy demands of the milking ewes without compromising their energy balance during the mid to late stages of lactation. Although high starch diets favor fat deposition against milk production, causing an increase of BCS in mid-lactating animals [46,52,53], this was not the case in our study. A possible explanation could be the regular modification of feedstuff quantities to efficiently meet, but not exceed, the nutritional demands of the ewes according to the stage of lactation and productivity.

The studied milk physical properties did not differ between the two groups, except with EC, which was significantly lower in group E. Although an increased EC has been associated with subclinical mastitis and increased SCC in dairy cows and small ruminants [52,53], in our study, we cannot conclude the potential improved mammary gland health status in the group E ewes, since SCC and TBC values were not significantly lower in that group. Moreover, EC is affected by the milk chemical composition, among other physiological factors, such as the parity number, lactation stage, and animal breed [54]. Specifically, it is negatively correlated with the milk fat content in dairy cows and small ruminants due to the nonconductive properties of fat globules [54,55], possibly explaining this decrease in the milk EC in the group E ewes, which presented significantly increased daily and total milk fat yields compared to the group C ewes. Nevertheless, this is an assumption which needs to be further Investigated to reveal the underlying mechanisms justifying the variation of milk EC. For this reason, a larger-scale assessment of health indicators of sheep fed with the two diets is necessary to conclude the effects of the two studied diets on the animal health and milk hygiene status.

Despite the encouraging results regarding the use of alternative protein sources, the gradual replacement of SBM in the diets of dairy sheep prerequisites addressing some limiting factors. In Greece and in many other European countries, the lack of experience and expertise in the cultivation of alternative protein crops (fava beans, rapeseed meal, lupin etc.) results in a low crop performance and the inadequate standardization of qualitative traits (protein, fat, moisture content, foreign material, etc.). In our study, the cost of the experimental ration was from EU 1.0 to EU 1.5/kg less than the control ration, despite the observed shortage of fava bean and the consequent increase in its price in the year of the study. In any case, the large-scale use of alternative protein-rich feedstuffs requires the extended cultivation of the respective crops to maintain a competitive cost against conventional rations (with SBM) and to succeed in sustainable production.

5. Conclusions

The partial replacement of SBM, used as the main protein source in a typical commercial concentrate ration for dairy sheep, with a mixture of rapeseed meal, cottonseed cake, and fava beans did not adversely affect the milk yield and any of the studied milk quality traits; on the contrary, it was associated with a favorable effect on daily and 100-day fat yields and on milk EC. The increase in the milk fat yield of ewes fed the experimental ration is possibly related to its greater NDF content. On the other hand, the decrease of EC in the experimental ration group may be linked to the increased milk fat content. However, the potential mechanisms justifying these findings need to be further investigated, assessing, at the same time, the nitrogen utilization, nutrient digestibility, and ruminal metabolism of the control and experimental rations. Although the results of our study support the efficient use of alternative protein sources for the substitution of SBM in the diets of intensively reared dairy sheep, further studies are deemed crucial for the assessment of the total replacement of SBM in terms of the animal health and productivity statuses and the overall farm sustainability.

Author Contributions: Conceptualization, A.I.G., G.T., I.P. and I.B.; methodology, A.I.K. and A.I.G.; formal analysis, A.I.K., T.L. and A.I.G.; investigation, A.I.K., M.M., F.C. and A.I.G.; resources, A.I.G.; data curation, A.I.K. and A.I.G.; writing—original draft preparation, A.I.K.; writing—review and editing, A.I.G., T.L., G.T., I.P. and I.B.; supervision, A.I.G.; project administration, A.I.G.; funding acquisition, A.I.G. All authors have read and agreed to the published version of the manuscript.

Funding: This research was funded by the Agricultural Cooperative 'Agrinio Union' within the Regional Operational Program 'Western Greece 2014-2020-RIS3', Project code: MIS-5040306, Project name: Feed for milk.

Institutional Review Board Statement: The study was conducted in compliance with the national animal welfare regulations i.e., the Presidential Decree 56/13 "Bringing Greek legislation into line with Directive 2010/63/EC of the European Parliament and of the Council of 22 September 2010 (L 276/33/20.10.2010) regarding the protection of animals used for experimental and other scientific purposes". The applied diagnostic veterinary procedures are not within the context of relevant EU legislation for animal experimentations (Directive 86/609/EC) and may be performed in order to diagnose animal diseases and improve animal welfare. In the current study, milk yield, body condition scoring, and milk sampling were performed as routine farm practices during milking of the animals.

Data Availability Statement: Not applicable.

Conflicts of Interest: The authors declare no conflict of interest.

References

1. Park, Y.W.; Haenlein, G.F.W. *Handbook of Milk of Non-Bovine Mammals (Google eBook)*; John Wiley & Sons: Hoboken, NJ, USA, 2008; ISBN 0470999721.
2. Haenlein, G.F.W. About the evolution of goat and sheep milk production. *Small Rumin. Res.* **2007**, *68*, 3–6. [CrossRef]
3. Molle, G.; Decandia, M.; Cabiddu, A.; Landau, S.Y.; Cannas, A. An update on the nutrition of dairy sheep grazing Mediterranean pastures. *Small Rumin. Res.* **2008**, *77*, 93–112. [CrossRef]
4. Kim, S.W.; Less, J.F.; Wang, L.; Yan, T.; Kiron, V.; Kaushik, S.J.; Lei, X.G. Meeting Global Feed Protein Demand: Challenge, Opportunity, and Strategy. *Annu. Rev. Anim. Biosci.* **2019**, *7*, 221–243. [CrossRef]
5. Ibáñez, M.A.; de Blas, C.; Cámara, L.; Mateos, G.G. Chemical composition, protein quality and nutritive value of commercial soybean meals produced from beans from different countries: A meta-analytical study. *Anim. Feed Sci. Technol.* **2020**, *267*, 114531. [CrossRef]
6. Boerema, A.; Peeters, A.; Swolfs, S.; Vandevenne, F.; Jacobs, S.; Staes, J.; Meire, P. Soybean trade: Balancing environmental and socio-economic impacts of an intercontinental market. *PLoS ONE* **2016**, *11*, e0155222. [CrossRef]
7. De Visser, C.L.M.; Schreuder, R.; Stoddard, F. The EU's dependency on soya bean import for the animal feed industry and potential for EU produced alternatives. *OCL-Oilseeds Fats* **2014**, *21*. [CrossRef]
8. Nemecek, T.; von Richthofen, J.S.; Dubois, G.; Casta, P.; Charles, R.; Pahl, H. Environmental impacts of introducing grain legumes into European crop rotations. *Eur. J. Agron.* **2008**, *28*, 380–393. [CrossRef]
9. Renna, M.; Cornale, P.; Lussiana, C.; Malfatto, V.; Fortina, R.; Mimosi, A.; Battaglini, L.M. Use of *Pisum sativum* (L.) as alternative protein resource in diets for dairy sheep: Effects on milk yield, gross composition and fatty acid profile. *Small Rumin. Res.* **2012**, *102*, 142–150. [CrossRef]
10. Tang, J.W.; Sun, H.; Yao, X.H.; Wu, Y.F.; Wang, X.; Feng, J. Effects of replacement of soybean meal by fermented cottonseed meal on growth performance, serum biochemical parameters and immune function of yellow-feathered broilers. *Asian-Australas. J. Anim. Sci.* **2012**, *25*, 393–400. [CrossRef]
11. Alves, F.J.L.; Ferreira, M.D.A.; Urbano, S.A.; de Andrade, R.D.P.X.; da Silva, Á.E.M.; de Siqueira, M.C.B.; de Oliveira, J.P.F.; Silva, J.D.L. Performance of lambs fed alternative protein sources to soybean meal. *Rev. Bras. Zootec.* **2016**, *45*, 145–150. [CrossRef]

12. Mordenti, A.L.; Merendi, F.; Fustini, M.; Formigoni, A. Effects of different protein plants in cows diet on milk for Parmigiano Reggiano production. *Ital. J. Anim. Sci.* **2007**, *6*, 463–465. [CrossRef]
13. Volpelli, L.A.; Comellini, M.; Gozzi, M.; Masoero, F.; Moschini, M. Pea (*Pisum sativum*) and faba beans (*Vicia faba*) in dairy cow diet: Effect on milk production and quality. *Ital. J. Anim. Sci.* **2012**, *11*, 217–222. [CrossRef]
14. Selmi, H.; Kamoun, M.; Tibaoui, G.; Ben Gara, A.; Rouissi, H. Effects of replacing corn and soya beans with white sorghum and faba beans on milk quality of Sicilo Sarde dairy ewes in Tunisia. *Options Méditerranéennes Série A Mediterr. Semin.* **2013**, *107*, 213–218.
15. Vasta, V.; Nudda, A.; Cannas, A.; Lanza, M.; Priolo, A. Alternative feed resources and their effects on the quality of meat and milk from small ruminants. *Anim. Feed Sci. Technol.* **2008**, *147*, 223–246. [CrossRef]
16. Silva, R.V.M.M.; de Carvalho, G.G.P.; Pires, A.J.V.; Pereira, M.L.A.; Pereira, L.; Campos, F.S.; Perazzo, A.F.; de Araújo, M.L.G.M.L.; de Oliveira Nascimento, C.; Santos, S.A.; et al. Cottonseed cake in substitution of soybean meal in diets for finishing lambs. *Small Rumin. Res.* **2016**, *137*, 183–188. [CrossRef]
17. Cavallini, D.; Mammi, L.M.E.; Biagi, G.; Fusaro, I.; Giammarco, M.; Formigoni, A.; Palmonari, A. Effects of 00-rapeseed meal inclusion in Parmigiano Reggiano hay-based ration on dairy cows' production, reticular pH and fibre digestibility. *Ital. J. Anim. Sci.* **2021**, *20*, 295–303. [CrossRef]
18. Sobotka, W.; Fiedorowicz-Szatkowska, E. The Effect of Replacing Genetically Modified Soybean Meal with 00-Rapeseed Meal, Faba Bean and Yellow Lupine in Grower-Finisher Diets on Nutrient Digestibility, Nitrogen Retention, Selected Blood Biochemical Parameters and Fattening Performance of Pigs. *Animals* **2021**, *11*, 960. [CrossRef]
19. Zagorakis, K.; Liamadis, D.; Milis, C.; Dotas, V.; Dotas, D. Effects of replacing soybean meal with alternative sources of protein on nutrient digestibility and energy value of sheep diets. *S. Afr. J. Anim. Sci.* **2018**, *48*, 489–496. [CrossRef]
20. Adewole, D.I.; Rogiewicz, A.; Dyck, B.; Slominski, B.A. Chemical and nutritive characteristics of canola meal from Canadian processing facilities. *Anim. Feed Sci. Technol.* **2016**, *222*, 17–30. [CrossRef]
21. Zhang, W.J.; Xu, Z.R.; Pan, X.L.; Yan, X.H.; Wang, Y.B. Advances in gossypol toxicity and processing effects of whole cottonseed in dairy cows feeding. *Livest. Sci.* **2007**, *111*, 1–9. [CrossRef]
22. Dalle Zotte, A.; Brand, T.S.; Hoffman, L.C.; Schoon, K.; Cullere, M.; Swart, R. Effect of cottonseed oilcake inclusion on ostrich growth performance and meat chemical composition. *Meat Sci.* **2013**, *93*, 194–200. [CrossRef] [PubMed]
23. Pelagalli, A.; Musco, N.; Trotta, N.; Cutrignelli, M.I.; Di Francia, A.; Infascelli, F.; Tudisco, R.; Lombardi, P.; Vastolo, A.; Calabrò, S. Chemical characterisation and in vitro gas production kinetics of eight faba bean varieties. *Animals* **2020**, *10*, 398. [CrossRef]
24. Kudlinskienė, I.; Gružauskas, R.; Daukšienė, A.; Dovidaitienė, G.; Želvytė, R.; Monkevičienė, I.; Šlyžius, E.; Urbšienė, D.; Racevičiūtė-Stupelienė, A.; Ots, M.; et al. Effect of extrusion on the chemical composition of the faba beans and its influence on lactation performance of dairy cows. *Zemdirbyste* **2020**, *107*, 87–94. [CrossRef]
25. Halmemies-Beauchet-Filleau, A.; Rinne, M.; Lamminen, M.; Mapato, C.; Ampapon, T.; Wanapat, M.; Vanhatalo, A. Review: Alternative and novel feeds for ruminants: Nutritive value, product quality and environmental aspects. *Animal* **2018**, *12*, S295–S309. [CrossRef]
26. Crépon, K.; Marget, P.; Peyronnet, C.; Carrouée, B.; Arese, P.; Duc, G. Nutritional value of faba bean (*Vicia faba* L.) seeds for feed and food. *Field Crops Res.* **2010**, *115*, 329–339. [CrossRef]
27. Heuzé, V.; Tran, G.; Delagarde, R.; Lessire, M.; Lebas, F. Faba bean (*Vicia faba*). Feedipedia, a Programme by INRAE, CIRAD, AFZ and FAO. Available online: https://www.feedipedia.org/node/4926 (accessed on 25 October 2021).
28. ICAR. *Section 16 Guidelines for Performance Recording in Dairy Sheep and Dairy Goats*; ICAR: Rome, Italy, 2018.
29. Russel, A.J.F.; Doney, J.M.; Gunn, R.G. Subjective assessment of body fat in live sheep. *J. Agric. Sci.* **1969**, *72*, 451–454. [CrossRef]
30. Wang, C.; Liu, J.X.; Zhai, S.W.; Lai, J.L.; Wu, Y.M. Effects of rumen-degradable-protein to rumen-undegradable-protein ratio on nitrogen conversion of lactating dairy cows. *Acta Agric. Scand. A Anim. Sci.* **2008**, *58*, 100–103. [CrossRef]
31. Brito, A.F.; Broderick, G.A. Effects of different protein supplements on milk production and nutrient utilization in lactating dairy cows. *J. Dairy Sci.* **2007**, *90*, 1816–1827. [CrossRef] [PubMed]
32. Huhtanen, P.; Hetta, M.; Swensson, C. Evaluation of canola meal as a protein supplement for dairy cows: A review and a meta-analysis. *Can. J. Anim. Sci.* **2011**, *91*, 529–543. [CrossRef]
33. Martineau, R.; Ouellet, D.R.; Lapierre, H. Feeding canola meal to dairy cows: A meta-analysis on lactational responses. *J. Dairy Sci.* **2013**, *96*, 1701–1714. [CrossRef]
34. Broderick, G.A.; Faciola, A.P.; Armentano, L.E. Replacing dietary soybean meal with canola meal improves production and efficiency of lactating dairy cows. *J. Dairy Sci.* **2015**, *98*, 5672–5687. [CrossRef] [PubMed]
35. Gidlund, H.; Hetta, M.; Krizsan, S.J.; Lemosquet, S.; Huhtanen, P. Effects of soybean meal or canola meal on milk production and methane emissions in lactating dairy cows fed grass silage-based diets. *J. Dairy Sci.* **2015**, *98*, 8093–8106. [CrossRef] [PubMed]
36. Maxin, G.; Ouellet, D.R.; Lapierre, H. Effect of substitution of soybean meal by canola meal or distillers grains in dairy rations on amino acid and glucose availability. *J. Dairy Sci.* **2013**, *96*, 7806–7817. [CrossRef]
37. Fombad, R.B.; Bryant, M.J. An evaluation of the use of cottonseed cake in the diet of growing pigs. *Trop. Anim. Health Prod.* **2004**, *36*, 295–305. [CrossRef]
38. Yehudi Coura de Assis, D.D.; Pinto de Carvalho, D.G.G.; Mauro Santos, D.E.; Almeida de Oliveira, D.F.; Garcia Melo Lopes de Araújo, D.M.L.; dos Santos Pina, D.D.; Alvarenga Santos, D.S.; Marta de Almeida Rufino, D.L. Cottonseed cake as a substitute of soybean meal for goat kids. *Ital. J. Anim. Sci.* **2019**, *18*, 124–133. [CrossRef]

39. Liponi, G.B.; Casini, L.; Martini, M.; Gatta, D. Faba bean (*Vicia faba* minor) and pea seeds (*Pisum sativum*) as protein sources in lactating ewes' diets. *Ital. J. Anim. Sci.* **2007**, *6*, 309–311. [CrossRef]
40. Bonanno, A.; Di Grigoli, A.; Vitale, F.; Alabiso, M.; Giosuè, C.; Mazza, F.; Todaro, M. Legume grain-based supplements in dairy sheep diet: Effects on milk yield, composition and fatty acid profile. *Anim. Prod. Sci.* **2016**, *56*, 130–140. [CrossRef]
41. Sanz Sampelayo, M.R.; Pérez, M.L.; Gil Extremera, F.; Boza, J.J.; Boza, J. Use of different dietary protein sources for lactating goats: Milk production and composition as functions of protein degradability and amino acid composition. *J. Dairy Sci.* **1999**, *82*, 555–565. [CrossRef]
42. Morales, E.R.; Alcaide, E.M.; Sampelayo, M.S. Milk production of dairy goats fed diets with different legume seeds: Effects of amino acid composition of the rumen undegradable protein fraction. *J. Sci. Food Agric.* **2008**, *88*, 2340–2349. [CrossRef]
43. Volpelli, L.A.; Comellini, M.; Masoero, F.; Moschini, M.; Lo Fiego, D.P.; Scipioni, R. Faba beans (*Vicia faba*) in dairy cow diet: Effect on milk production and quality. *Ital. J. Anim. Sci.* **2009**, *9*, 138–144. [CrossRef]
44. Tufarelli, V.; Khan, R.U.; Laudadio, V. Evaluating the suitability of field beans as a substitute for soybean meal in early-lactating dairy cow: Production and metabolic responses. *Anim. Sci. J.* **2012**, *83*, 136–140. [CrossRef] [PubMed]
45. Selmi, H.; Bahri, A.; Rouissi, H. Nutrition for Lactation of Dairy Sheep. In *Lactation in Farm Animals—Biology, Physiological Basis, Nutritional Requirements, and Modelization*; IntechOpen: London, UK, 2020; pp. 1–12. [CrossRef]
46. Lunesu, M.F.; Decandia, M.; Molle, G.; Atzori, A.S.; Bomboi, G.C.; Cannas, A. Dietary starch concentration affects dairy sheep and goat performances differently during mid-lactation. *Animals* **2021**, *11*, 1222. [CrossRef] [PubMed]
47. Pulina, G.; Nudda, A.; Battacone, G.; Cannas, A. Effects of nutrition on the contents of fat, protein, somatic cells, aromatic compounds, and undesirable substances in sheep milk. *Anim. Feed Sci. Technol.* **2006**, *131*, 255–291. [CrossRef]
48. Bencini, R.; Stanislao Atzori, A.; Nudda, A.; Battacone, G.; Pulina, G. *Improving the Quality and Safety of Sheep Milk*; Woodhead Publishing Limited: Cambridge, UK, 2010; ISBN 9781845698065.
49. Bauman, D.E.; McGuire, M.A.; Harvatine, K.J. Mammary Gland, Milk Biosynthesis and Secretion: Milk Fat. *Encycl. Dairy Sci. Second Ed.* **2011**, *1*, 352–358. [CrossRef]
50. Nudda, A.; Battacone, G.; Neto, O.B.; Cannas, A.; Helena, A.; Francesconi, D.; Atzori, A.S.; Pulina, G. Invited Review Feeding strategies to design the fatty acid profile of sheep milk and cheese. *Rev. Bras. Zootec.* **2014**, *43*, 445–456. [CrossRef]
51. Kenyon, P.R.; Maloney, S.K.; Blache, D. Review of sheep body condition score in relation to production characteristics. *N. Zeal. J. Agric. Res.* **2014**, *57*, 38–64. [CrossRef]
52. Caria, M.; Chessa, G.; Murgia, L.; Todde, G.; Pazzona, A. Development and test of a portable device to monitor the health status of Sarda breed sheep by the measurement of the milk electrical conductivity. *Ital. J. Anim. Sci.* **2016**, *15*, 275–282. [CrossRef]
53. Norberg, E.; Hogeveen, H.; Korsgaard, I.R.; Friggens, N.C.; Sloth, K.H.M.N.; Løvendahl, P. Electrical conductivity of milk: Ability to predict mastitis status. *J. Dairy Sci.* **2004**, *87*, 1099–1107. [CrossRef]
54. Mabrook, M.F.; Petty, M.C. Effect of composition on the electrical conductance of milk. *J. Food Eng.* **2003**, *60*, 321–325. [CrossRef]
55. Romero, G.; Roca, A.; Alejandro, M.; Muelas, R.; Díaz, J.R. Relationship of mammary gland health status and other noninfectious factors with electrical conductivity of milk in Manchega ewes. *J. Dairy Sci.* **2017**, *100*, 1555–1567. [CrossRef]

animals

MDPI

Article

The Effect of Herbal Feed Additives in the Diet of Dairy Goats on Intestinal Lactic Acid Bacteria (LAB) Count

Joanna Foksowicz-Flaczyk [1], Jacek Antoni Wójtowski [2,*], Romualda Danków [3], Przemysław Mikołajczak [4,5], Jan Pikul [3], Agnieszka Gryszczyńska [4], Zdzisław Łowicki [4], Karolina Zajączek [4] and Daniel Stanisławski [6]

1 Department of Innovative Biomaterials and Nanotechnologies, Institute of Natural Fibres and Medicinal Plants, Wojska Polskiego 71b, 60-630 Poznań, Poland; joanna.flaczyk@iwnirz.pl
2 Department of Animal Breeding and Product Quality Assessment, Faculty of Veterinary Medicine and Animal Science, Poznań University of Life Science, Złotniki, ul. Słoneczna 1, 62-002 Suchy Las, Poland
3 Department of Dairy and Process Engineering, Faculty of Food Science and Nutrition, Poznań University of Life Sciences, ul. Wojska Polskiego 31/33, 60-624 Poznań, Poland; romualda.dankow@up.poznan.pl (R.D.); jan.pikul@up.poznan.pl (J.P.)
4 Department of Pharmacology and Phytochemistry, Institute of Natural Fibres and Medicinal Plants, Kolejowa 2, 62-064 Plewiska, Poland; przemmik@ump.edu.pl (P.M.); agnieszka.gryszczynska@iwnirz.pl (A.G.); zdzislaw.lowicki@iwnirz.pl (Z.Ł.); karolina.zajaczek@iwnirz.pl (K.Z.)
5 Department of Pharmacology, Poznań University of Medical Sciences, ul. Rokietnicka 5a, 60-806 Poznań, Poland
6 Computer Laboratory, Poznań University of Life Sciences, ul. Wołyńska 33, 60-637 Poznań, Poland; daniel.stanislawski@up.poznan.pl
* Correspondence: jacek.wojtowski@up.poznan.pl

Citation: Foksowicz-Flaczyk, J.; Wójtowski, J.A.; Danków, R.; Mikołajczak, P.; Pikul, J.; Gryszczyńska, A.; Łowicki, Z.; Zajączek, K.; Stanisławski, D. The Effect of Herbal Feed Additives in the Diet of Dairy Goats on Intestinal Lactic Acid Bacteria (LAB) Count. *Animals* **2022**, *12*, 255. https://doi.org/10.3390/ani12030255

Academic Editors: Panagiotis Simitzis and Athanasios I. Gelasakis

Received: 29 November 2021
Accepted: 18 January 2022
Published: 21 January 2022

Simple Summary: The prohibition on the use of antibiotics in animal nutrition has resulted in the more frequent use of phytobiotics, which are natural medical preparations made from herbs. When used in the nutrition of ruminants, phytobiotic preparations affect the motility of the gastrointestinal tract and the secretion of digestive juices, and also stimulate the development of the intestinal microbiota. Their effect on the development of lactic acid bacteria (LAB), with subsequent effects on the degree of microbial homeostasis in the gastrointestinal tract, is particularly important. The aim of the present study was to evaluate the effects of herbal supplements on lactic acid bacteria (LAB) count in the faeces of lactating dairy goats. It was assumed that the specific chemical composition of herbal supplements would positively affect the digestive processes of does, and thus the growth of lactic acid bacteria (LAB) colonies. The research was conducted on dairy goats assigned to five nutrition groups of twelve animals each. The animals in the experimental groups received a supplement made of (seven or nine) herbs at a rate of 20 g or 40 g per animal per day. A statistically significant effect of lactation stage on the intestinal *Lactobacillus* bacteria count was found. The highest concentration of LAB was found in the group receiving a feed supplement consisting of nine herbs at 20 g per animal per day. A probiotic strain of *Lactobacillus fermentum* absent from the control goats was identified in the faecal samples of goats that receiving the herbal supplement.

Abstract: Sixty dairy goats of the Polish white improved breed were randomly assigned to five feeding groups of twelve animals each. The animals received a supplement containing seven herbs at 20 or 40 g/animal/day (experimental groups 1 and 2) and a supplement containing nine herbs at 20 or 40 g/animal/day (experimental groups 3 and 4)m, along with pelleted concentrate feed. Group 5 (the control group) received pelleted feed without any herbal supplements. A significant effect of herbal feed additive on lactic acid bacteria (LAB) count was observed ($p < 0.001$). The highest number density of LAB was found in the goats receiving the feed additive with nine herbs at 20 g/animal per day ($p < 0.05$). There was a statistically significant effect of lactation stage on intestinal LAB count ($p < 0.001$). Regardless of the feeding group, the highest number density of LAB was found in animals at the peak of lactation. The LAB count was also affected by the interaction of diet group × lactation stage ($p < 0.0001$). A probiotic strain of *Lactobacillus fermentum* was identified in the faecal samples of goats receiving the herbal additive, but not in the controls. Genetic identification of the microorganisms isolated from the faeces of the experimental goats did not reveal the presence of harmful mould spores, although spores of the fungus Aspergillus fumigatus were detected in the controls.

Keywords: herbal feed additives; intestinal lactic acid bacteria (LAB); dairy goats

1. Introduction

The prohibition on the use of antibiotics in animal nutrition has resulted in the increased use of natural substances derived from medicinal plants [1]. Herbs containing bioactive ingredients—phytobiotics—have a particularly broad spectrum of action [2,3].

Phytobiotic mixtures are produced from wild plants or extracted from field crops [4]. Herbal raw materials are those parts of plants in which the accumulation of active ingredients is relatively high, and may include leaves, rhizomes, roots, flowers, bark, fruit, or seeds. The stimulating or prophylactic and therapeutic properties of plants are determined by their bioactive ingredient content, which is maximized by harvesting at the optimal vegetative phase, the appropriate conditions and place of harvesting, proper drying, and storage [5]. Even under proper storage conditions, the properties intensity of the active ingredients of herbs diminishes over time [6]. Production waste from the herbal industry may also be used as a feed additive, provided that it still has an appropriate active ingredient content. One example is the endosperm of milk thistle, which is a waste product in the production of silymarin [7]. The most important groups of bioactive ingredients that are found in herbal raw materials are tannins, saponins, essential oils, flavonoids, glycosides, and alkaloids [8].

The phytobiotic preparations used in feeding domestic animals, especially ruminants, can enhance taste sensations and stimulate appetite. As regulators of digestive functions, they also affect gastrointestinal (GI) motility and the secretion of digestive juices, reduce the occurrence of diarrhoea, and regulate the pH of the GI tract [9]. Some of them may also have a protective effect (such as fenugreek and flax), regulate metabolism (e.g., knotgrass, nettle), or affect the quality of animal products (e.g., garlic and calendula flower) [10].

Animals' taste preferences should also be taken into consideration when formulating herbal mixes. Herbs usually contain high levels of essential oils and there may be problems with their uptake by some ruminants, such as sheep [11].

The herbal feed additives provided to ruminants stimulate digestive processes by supporting rumen microorganisms [9,12]. In particular, their effect on the growth of probiotic LAB is important, because it affects the degree of gastrointestinal (GI) microbial homeostasis. The gut flora balance constitutes an effective barrier against pathogen colonisation, influences the production of metabolic substrates (e.g., vitamins and short-chain fatty acids), and positively stimulates the immune system [13].

There are more than 180 species of *Lactobacillus*, and these include the homofermentative and mesophilic *Lactococcus lactis*, the best-known species of LAB. Strains with proven probiotic properties are considered most valuable, and include *Lactobacillus rhamnosus*, *Lactobacillus casei*, *Lactobacillus bulgaricus*, *Lactobacillus salivarius*, *Lactobacillus plantarum*, *Lactobacillus acidophilus* and *Lactobacillus helveticus* [14].

This study aimed to evaluate the effects of herbal supplements incorporated into the diets of lactating dairy goats on faecal LAB count. We assumed that the specific chemical composition of the herbal supplements would have a positive effect on the digestive processes of the animals, thus increasing the colonies of LAB that fortify microbial homeostasis in the GI tract.

2. Materials and Methods

2.1. Ethical Approval

All the research was performed in accordance with the Polish Act on the protection of animals used for scientific or teaching purposes, which complies with EU legislation on the protection of animals used for scientific purposes. All procedures were approved by the Local Bioethics Committee for Animal Testing (Poznań, Poland; decision no. 57/2020).

2.2. Location and Animal Material

The experiment was conducted on sixty Polish white improved goats kept on a specialised farm located in northwestern Poland (Bukowiec, 52°51′41″ N; 16°52′12″ E) in the Wielkopolska region. Clinically healthy goats were selected for the experiment. The somatic cell count (SCC) measured immediately prior to the experiment (during the third week of lactation) was at an acceptable level, and did not exceed 800×10^3/mL. The animals were aged 20–30 months, were in their second lactation, and had a body weight of 56–60 kg. The experiment started when goats were approximately 28.1 ± 2.7 days in milk (DIM).

The animals were randomly assigned to five feeding groups of twelve goats each:

Group 1 (receiving 20 g of herbal supplement—mix of seven herbs).
Group 2 (receiving 40 g of herbal supplement—mix of seven herbs).
Group 3 (receiving 20 g of herbal supplement—mix of nine herbs).
Group 2 (receiving 40 g of herbal supplement—mix of nine herbs).
Group 5 (control group, no herbal supplements).

The amount of herbal supplement provided (20 g or 40 g/goat/day) was based on the experiment of Jarzynowska and Peter [15] on dairy sheep.

Goats were tagged using electronic transponders and coloured collars, varied by group, with plastic numbers.

2.3. Herbal Supplements

The herbal supplements provided to the experimental goats were composed of seven herbs (herbal supplement I) or nine herbs (herbal supplement II). The choice of herbs for the supplements was established on the basis of our previous experiments (unpublished) and herbs in supplement for dairy sheep in the Jarzynowska and Peter's experiment [15].

Herbal mix 1 included common nettle *Urtica dioica* L. (herb); common agrimony *Agrimonia eupatoria* (herb—dried flowering shoot tips); caraway *Carum carvi* (fruit); coriander *Coriandrum sativum* (fruit); fenugreek *Trigonella foenum graecum* L. (seeds); plantain *Plantago lanceolata* L. (herb); and purple willow *Salix purpurea* (bark).

Herbal mix 2 contained different proportions of herbs to that used in herbal mix 1. The herbs included were common nettle *Urtica dioica* L. (herb); common agrimony *Agrimonia eupatoria* (herb—dried flowering shoot tips); coriander *Coriandrum sativum* (fruit); fenugreek *Trigonella foenum graecum* L. (seeds), as well as fennel *Foeniculum vulgare* (fruit); peppermint *Mentha piperita* (leaves); chamomile *Matricaria chamomilla* L. (flower clusters); milk thistle *Silybum marianum* (endosperm); and thyme *Thymus vulgaris* (leaves).

A detailed contribution of particular herbal components were included in these supplements is presented in our patent applications (Polish Patent Office submissions P.4334426 and P.433779)

2.4. Animal Nutrition

The diets were formulated to meet the animals' nutrient requirements: 2.12 UFM (unit for milk production) and 185 g PDI (protein truly digestible in the small intestine) to obtain an assumed milk yield of 3.0 kg and 3.8% of fat [16].

The ingredients (% DM) of diet offered to dairy goats were:

15.6% maize silage; 21.6% grass hay silage; 7.8% brewers' grain silage; 26.4% concentrate mixture; 10.3% meadow hay; 10.6% experimental concentrate (with herbal mix); 4.4% dried sugar beet pulp; and 3.3% barley straw.

The chemical composition of the diet was 451 g kg^{-1} of DM organic matter, 163 g kg^{-1} of DM crude protein, 267 g kg^{-1} of DM acid detergent fibre (ADF), and 401 g kg^{-1} of DM neutral detergent fibre (NDF).

All the ingredients, other than the experimental concentrate with the herbal mix, were part of the total mixed ration (TMR) feed and were offered to the animals once a day. Goats had free access to water and a mineral salt lick.

The herbal supplement was provided to the animals in the prepared pelleted concentrate feed (cereal grains, rapeseed meal, sunflower meal), containing the concentration of the mixes of seven herbs (Group 1 and Group 2) and nine herbs (Group 3 and Group 4). Group 5 was a control group and thus did not receive the herbal supplement.

Group 1 (G1): basal diet plus 20 g DM herbal mix 1 in 300 g of concentrate (herbal mix 1, 6.6 g of 100 g^{-1} concentrate dry matter);
Group 2 (G2): basal diet plus 40 g DM herbal mix 1 in 300 g of concentrate (herbal mix 1, 13.2 g of 100 g^{-1} concentrate dry matter);
Group 3 (G3): basal diet plus 20 g DM herbal mix 2 in 300 g of concentrate (herbal mix 1, 6.6 g of 100 g^{-1} concentrate dry matter);
Group 4 (G4): basal diet plus 40 g DM herbal mix 2 in 300 g of concentrate (herbal mix 1, 13.2 g of 100 g^{-1} concentrate dry matter);
Group 5 (CTRL): basal diet plus 300 g concentrate (no herbs; control group).

The composition of experimental concentrates is shown in Table 1.

Table 1. The composition of the experimental concentrates (% DM).

Item	Dietary Treatment		
	Groups 1 and 3	**Groups 2 and 4**	**Group 5 (Control)**
Ingredient (% DM)			
Wheat bran	17	13	17
Triticale	18.6	18	18.6
Rapeseed meal	17	16.5	17
Sunflower meal	10	9.5	10
Corn DDGS [a]	5	5	5
Rye	7	6	7
Wheat	5	5	5
Barley	4	4	4
Dried grasses	0	0	6.6
Herbs	6.6	13.2	0
Sugarcane molasses	2	2	2
Dried sugar beet pulp	4.2	4.2	4.2
Minerals and vitamins [b]	2.5	2.5	2.5
Fodder chalk	0.1	0.1	0.1
Salt	1	1	1
Composition (g kg^{-1} DM)			
Organic matter	927	926	928
Crude protein	229	223	224
Crude fat	36	33	34
Crude fibre	86	92	87

Groups 1 and 2: a mix of seven herbs; groups 3 and 4: a mix of nine herbs; group 5: control group (no herbal supplements); [a] corn DDGS, distiller's dried grain with solubles from the production of biodiesel and ethanol; and [b] 1 kg of minerals and vitamins contains 300,000 units of vitamin A, 30,000 units of vitamin D_3, 1.5 g of vitamin E, 0.5 g of Fe, 2.5 g of Zn, 65.0 g of Mg, 0.015 g of Co, 3.0 g of Mn, 0.01 g of I, 0.003 g of Se, 60 g of Na, 240 g of Ca, and 120 g of P.

2.5. Microbiological Tests of Faeces

The faeces underwent microbiological testing to determine the amount of LAB, in order to assess the effects of the herbs on the microbiota of the digestive tract of the dairy goats. The faeces for testing were collected from the animals of the five groups at four times: before the start of the experiment (T0) and at the end of the first (T1), second (T2), and third (T3) trimesters of lactation. Faecal samples were collected directly from the previously disinfected milking stall floor during morning milking of each animal. Faeces were collected in individually labelled, sterile, 50 mL plastic containers. All samples were placed in an ice thermostat and were transported to the laboratory within two hours. Each collected sample (10 g) were individually dissolved in 10 g of sterile saline and shaken using a vortex mixer for 1 h; they were then diluted using the decimal dilution method and plated on Petri plates. The Petri plates were flooded with MRS agar broth nutrient medium containing 20.0 g/L agar, 20.0 g/L glucose, 10.0 g/L peptone K, 8.0 g/L Lab-Lemco powder, 4.0 g/L yeast extract, 1 mL sorbitan monooleate, 2.0 g/L dipotassium hydrogen phosphate, 5.0 g/L sodium acetate, 2.0 g/L triammonium citrate, 0.2 g/L magnesium sulphate, and 0.05 g/L manganese sulphate. The Petri plates were placed in an incubator and incubated under anaerobic conditions at 35–37 °C for 48–72 h. After the incubation, the number of single bacterial colonies grown on the plates was determined.

Identification of LAB Strains

After incubation, a similar number of samples from all experimental dates were collected from both the aerobically and anaerobically cultured samples. The cultured LAB strains were then identified.

Genetic material (DNA) was isolated from the most frequently and morphologically repetitive LAB colonies. Identification of LAB strains involved the following stages:

(1) Isolation of DNA from colonies grown on the plates.

Twelve isolations of genetic material (DNA) were performed on colonies of microorganisms provided on plates. DNA was isolated using CHELEX (Bio-Rad, Hercules, CA, USA) with the addition of enzymes to digest the cell wall.

(2) Amplification of the 16SrRNA gene fragment using polymerase chain reaction (PCR) with specific primers and sequencing of PCR arrays.

To confirm the presence of bacteria in the sample, amplification by PCR of 16S rDNA query fragments was performed using specific primers:
27F: 5-AGAGTTTGATCMTGGCTCAG-3;
1492R: 5-GGTTACCTTGTTACGACTT-3;
on the DNA template isolated from the colony.

The amplification reaction was performed in the ABI 9700 thermocycler (Life Technologies, Waltham, MA, USA) using thermostable OptiTaq polymerase (Eurx, Gdansk, Poland).

PCR conditions:

(1) 95 °C for 3 min.
(2) 95 °C for 15 s.
(3) 55 °C for 15 s.
(4) 72 °C for 90 s.
(5) Steps 2–4 were repeated 30 times.
(6) 72 °C for 2 min.
(7) 10 °C until cooled.

All the samples proved positive for amplification. PCR products were then purified, and sequencing was performed using a BigDye Terminator Mix v3.1 kit (Applied Biosystems, Forest City, CA, USA), an ABI3730xl genetic analyser, and specific primers.

The reads (from the bacterial 16S-rDNA-specific primers 27F and 1492R) were assembled into contigs, yielding a consensus sequence.

(3) Amplification by PCR of the internal transcribed spacer (ITS) fragment using specific primers and sequencing of PCR arrays.

To determine if there were any fungi present in the sample, amplification by PCR of ITS fragments was performed using specific primers:

ITS1: 5-TCCGTAGGTGAACCTGCGG-3;
ITS4: 5-TCCTCCGCTTATTGATATGC-3;
on the DNA template isolated from the colony.

All samples proved positive for amplification. The PCR products were then purified, and sequencing was performed using the BigDye Terminator Mix v3.1 kit, ABI3730xl genetic analyser, and specific primers. The reads from the ITS primers (ITS1-F and ITS4-R) were assembled into appropriate contigs, yielding a consensus sequence.

(4) Alignment of the obtained sequences and the NCBI database.

The consensus sequences were compared with the NCBI database (GeneBank, https://www.ncbi.nlm.nih.gov/gene/?term=, accessed on 25 October 2017) using the BLAST software (NCBI, Bthesda, MD, USA).

2.6. Statistical Analysis

Data were analyzed using SAS version 9.4 (2014, SAS Institute, Cary, NC, USA). Before analysis was conducted, all the data were evaluated for normality using PROC UNIVARIATE SAS (SAS Institute, Cary, NC, USA). As no normal distribution was found in the collected samples (Figure 1), the count of LAB determined in the faeces underwent the Box–Cox transformation with an estimated $\lambda = -0.114851$.

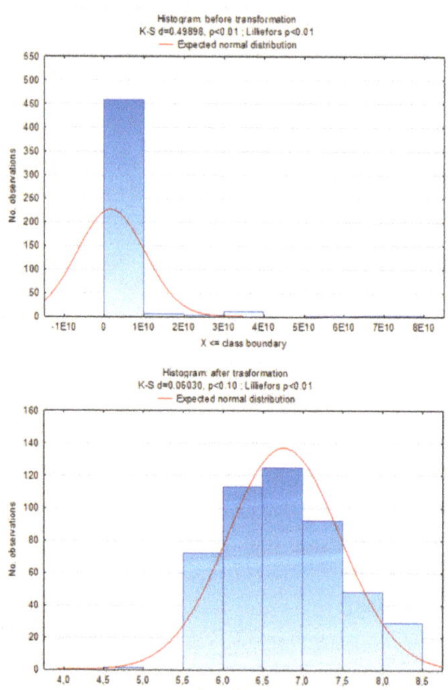

Figure 1. The distribution of LAB count before and after the Box–Cox transformation.

Data were analysed using a PROC MIXED model (version 9.4, SAS Institute, Cary, NC, USA). The lowest Akaike Information Criterion (AIC) was used to determine the appropriate within-subject covariance structure, and the compound symmetry (CS) was selected accordingly. Data were analysed as repeated measures (goat effect) using the following model: Yjk = μ + gi + tj + tgij + eijk, where: Yijk $Y_{jk} = \mu + g_i + t_j + tg_{ij}$ + eijk, where the Y_{ijk} are the observation means, μ is the overall mean, the g_i are the fixed effects of the groups (l = 1, 2, 3, 4, 5), the t_j are the fixed effects of the time of measurement (k = 1, 2, 3, 4), the tg_{ij} are the interaction of group × time, and the e_{ijk} are the residual errors.

When differences were detected in terms of treatment or interactions of treatment with time, separation of means was conducted using a Tukey's adjustment for the probability. The statistical significance was considered to be $p \leq 0.05$

3. Results

3.1. The Effects of Experimental Factors on LAB Count

The effects of the experimental factors on LAB count are shown in Tables 2 and 3. There was a highly significant effect of feeding group on LAB count (Table 2, $p < 0.001$). The LAB count was highest in the faeces of animals in Groups 3 and 2 ($p < 0.05$). The LAB counts of the faeces of goats in Groups 1, 2, and 4 were similar ($p < 0.05$). Excluding Group 4, the LAB content was significantly higher than in controls ($p < 0.05$). There was a highly significant effect of sampling time on LAB count ($p < 0.0001$, Table 2). The introduction of herbs into the diet of dairy goats increased the LAB count in the GI tract. There was a significant increase in the LAB count in stages T1 and T2 of lactation. The LAB count was highest in T2, the peak period of lactation, which indicates that there was significant effect of the stage of lactation on LAB count (Table 3). The LAB count determined in the faeces of Group 3 during the T2 measurement period was significantly higher than that in all other animal groups.

Table 2. Effects of experimental factors on LAB count.

LAB	Group					Time				SE	Group	Time	Group × Time
	1	2	3	4	5	T0	T1	T2	T3		*p*-Value		
Transformed	6.82 [a]	6.93 [ab]	7.03 [b]	6.75 [ac]	6.60 [c]	6.35 [a]	6.80 [b]	7.24 [c]	6.91 [b]	0.03	0.0001	0.0001	0.0033
CFU	3.15×10^8	2.37×10^9	1.92×10^9	1.95×10^9	6.92×10^5	9.40×10^4	7.03×10^8	4.54×10^9	4.97×10^6	-	-	-	-

Means marked with different letters are statistically different at $p \leq 0.05$. Transformed: value after Box–Cox transformation (first row); CFU: number of colony-forming units (second row); and SE: standard error.

Table 3. Effects of measurement time on LAB count in groups.

LAB	Group	Time				SE
		T0	T1	T2	T3	
Transformed	1	6.35 [a]	6.82 [b]	7.34 [*cef*]	6.77 [b]	0.07
CFU		9.33×10^4	1.51×10^6	1.25×10^9	2.58×10^6	-
Transformed	2	6.37 [a]	7.01 [b]	7.35 [*bef*]	6.98 [b]	0.08
CFU		9.41×10^4	2.36×10^9	7.13×10^9	4.55×10^6	-
Transformed	3	6.36 [a]	6.98 [b]	7.72 [*ce*]	7.07 [b]	0.09
CFU		9.51×10^4	4.28×10^7	7.63×10^9	1.19×10^7	-
Transformed	4	6.33 [a]	6.65 [ab]	7.11 [*bfg*]	6.90 [b]	0.08
CFU		9.33×10^4	1.11×10^9	6.69×10^9	4.03×10^6	-
Transformed	CTRL	6.36 [a]	6.55 [ab]	6.66 [*abg*]	6.81 [b]	0.05
CFU		9.41×10^4	5.39×10^5	4.25×10^5	1.71×10^6	-

Means marked with different letters are statistically different at $p \leq 0.05$. Reading across rows, the letters mark the significance of differences within experimental groups. Reading down columns, the letters mark the significance of differences among experimental groups by time of measurement (*italics*). Transformed: value after Box–Cox transformation (first row); CFU: number of colony-forming units (second row); and SE: standard error.

3.2. Identification of LAB Strains

Table 4 shows the results of identifying microorganisms from the DNA of the most frequently and morphologically repeated bacterial colonies.

Table 4. Alignment of consensus sequence for the 16S query fragments with the subject sequence in samples collected from animals of the experimental and control groups.

Species of Bacteria	Similarity	Sequence Coverage
	Experimental and control groups	
Lactobacillus buchneri strain JCM 1115	99.8%	100%
Enterococcus faecium strain ATCC 19434	100%	100%
Enterococcus mundtii strain NBRC 100490	100%	100%
	Experimental group	
Lactobacillus fermentum strain NBRC 15885	99.9%	100%
	Control group	
Aspergillus fumigatus isolate C1946	100%	100%

The following species of microorganism were found in the genetic material isolated from the experimental and control samples: *Lactobacillus buchneri* strain JCM 1115, *Enterococcus faecium* strain ATCC 19434, and *Enterococcus mundtii* strain NBRC 100490. Furthermore, *Lactobacillus fermentum* strain NBRC 15885 was present in faecal samples collected from the goats in the experimental groups.

The genetic identification of faecal samples of the controls revealed, in addition to LAB, the presence of spores of the fungus *Aspergillus fumigatus* in the digestive tract of goats. This is a pathogenic exogenous fungal species that may cause various infections in animals. In ruminants, in addition to weakening of the immune system, this species can affect the throat, nasal mucous membranes, or lungs, and can cause acute enteritis.

No mould spores were detected in the samples from the goats fed with the herb supplements.

4. Discussion

LAB are gram-positive, nonsporulating bacteria with low guanine-cytosine (GC) pairs in the genome. This group was singled out for its ability to perform carbohydrate fermentation with production of lactic acid, rather than for its phylogenetic relationships [17]. Although most LAB are anaerobia, some species may tolerate low levels of oxygen. LAB have strong auxotrophy, and are thus found in environments that accommodate their high nutritional requirements—i.e., that are rich in amino acids, purines, and pyrimidines. LAB can be found in milk and its derivative products, and are also components of the physiological flora (microbiota) of mammals. This group of microorganisms includes species of the *Lactococcus, Streptococcus, Pediococcus, Leuconostoc,* and *Lactobacillus* genera. Probiotic species are particularly valuable LAB [18]. These bacteria modulate the gut flora and thus maintain its homeostasis. They provide protection against pathogenic bacteria by competing with them for colonised surface. LAB can secrete compounds that inhibit pathogen growth (lactic acid, short-chain fatty acids, hydrogen peroxide, and substances that act as bacteriocins) [17]. Moreover, they stimulate the immune system and reduce the risk of allergic reactions [17].

Probiotic strains play the most significant role in supporting the treatment of GI diseases—especially viral diarrhoea and antibiotic-associated diarrhoea—and autoimmune disorders [14,17]. Their positive effects on metabolic diseases (hyperlipidaemia, diabetes, and obesity) have also been observed [16]. Probiotics are also credited with alleviating symptoms of lactose intolerance, increasing intestinal absorption of nutrients, lowering cholesterol level, improving intestinal peristalsis, and decreasing the activity of enzymes associated with carcinogenesis [14].

The literature contains few results from research works concerning the microbial composition of faeces in ruminant animals, mostly focusing on dairy cattle. Experiments conducted on calves have shown the relationship between faecal microbiota and age [19], nutritional diet [20,21], antibiotic therapy [22–24], and calf health status [25].

The effect of limit-feeding diets with different forage-to-concentrate ratios on faecal bacterial community composition in Holstein heifers has been studied by Zhang et al. [26] and others. In the study of Kim et al. [27], concerning bacterial diversity in the faeces of cattle fed different diets, faecal samples were collected from cattle fed a finishing steer diet ("moderate grain diet"), a late growing diet ("high-grain diet"), and from heifers fed an early growing diet ("silage/forage"). The taxonomic composition of faecal microbiota in these three diet groups was compared based on the mean of the relative abundance (reads of taxon divided by total number of reads in the sample). The abundance of *Lactobacillus* was different ($p < 0.001$) in the three groups; the high-grain diet group had the greatest abundance (1.50% of total sequences).

There have been few reports on the effect of lactation stages in ruminants on faecal microbiota composition. The report of Huang et al. [28] shows a significant effect of lactation period on diversity at the phylum level in the faecal bacterial community. This means that lactation stages induce a variation in the faecal bacterial community [28].

There are very few results concerning the composition of the gut microbiota of goats. The study of Draksler et al. [29] is one of the few reports to describe the number density of LAB. The LAB content, identified in faecal samples of Creole goats kept in northwestern Argentina, reached its highest value in the first two weeks of a goat kid's life and ranged from 5.58 to 7.15 $\log_{10}$ units/g of faeces [29]. In animals aged 30–60 days the CFU of LAB count decreased, reaching 5.24 and 5.43 $\log_{10}$ units/g of faeces, respectively. LAB content held stable from ninety days of age onwards. For animals in each age range, the $\log_{10}$ value of LAB was 4.61 (90 days), 4.93 (120 days), 4.82 (150 days), 4.52 (180 days), and 4.52 (270 days) [29].

Stella et al. [30] evaluated the effect of administering live *Saccharomyces cerevisiae* on milk production, milk composition, blood metabolites, and faecal flora in early lactating dairy goats. There was a significant effect on faecal flora. The differences between the control and experimental groups in terms of colony counts of *Lactobacilli* were particularly pronounced and statistically significant at sixty and ninety days of lactation, at 5.05 versus 6.21 and 4.89 versus 6.37 $\log_{10}$/g of faeces, respectively.

The following LAB species were identified in faecal samples: *Lactobacillus buchneri* strain JCM 1115 (experimental and control groups) and, only in the experimental group samples, *Lactobacillus fermentum* strain NBRC 15885. Both L. *buchneri* and L. *fermentum* are typical probiotic bacteria with proven antioxidant activities. According to the experiment conducted by Shokryazdan et al. [31], these strains had good antimicrobial activity against selected pathogenic strains of humans and exhibited stronger antimicrobial activity than the reference strain, L. *casei* Shirota.

The genetic identification performed as part of our experiment also revealed the presence of microbes such as *Enterococcus faecium* strain ATCC 19434 and *Enterococcus faecium* strain NBRC 100490. *Enterococcus* is a genus that is commonly found in the gut microbiota of ruminant animals, especially in the first stages of life that are not related to rumination. Jiao et al. [32], who studied the gut microbiota of goat kids during their first week of life, estimated the proportion of *Enterococcus* in the total species composition of the gut microbiota at 30.94% of the sequences under study.

It should be noted that spores of an exogenous fungus of the pathogenic species *Aspergillus fumigatus* were identified in goat faeces collected from controls. In ruminant animals this species may lead to various infections and even acute enteritis under extreme conditions [32].

5. Conclusions

There is a significant effect of the herbal feed additive on LAB count ($p < 0.001$). The highest number density of LAB was found in the group of goats receiving a feed additive that contained nine herbs at 20 g/animal per day ($p < 0.05$).

There was a statistically strong effect of lactation stage on intestinal LAB count ($p < 0.001$) The greatest number density of LAB was found in animals of all feeding groups at the peak of lactation (T2). Moreover, there is a highly significant interaction of feeding group $\times$ time of faecal sample collection ($p < 0.0001$).

The valuable probiotic species *Lactobacillus fermentum* strain NBRC 15885, is present in faecal samples of goats receiving a herbal additive compared to controls. The results of the genetic identification of faecal samples collected from the animals receiving the herbal supplement did not reveal the presence of mould spores, which are potentially harmful to the health of small ruminants; however, these spores were identified in controls.

6. Patents

A patent for "Herbal feed additives and their application" was submitted to the Polish Patent Office for intellectual protection of this technology (Polish Patent Application No. P.433779 and P.434426).

Author Contributions: Conceptualization: J.A.W. and J.F-F.; methodology: J.A.W. and J.F.-F.; software: D.S.; validation: J.A.W., J.F.-F., P.M., R.D., and J.P.; formal analysis: J.A.W., J.F.-F., R.D., and D.S.; investigation: J.A.W., J.F.-F., R.D., and A.G.; resources: J.A.W., J.F.-F., R.D., A.G., Z.Ł., and K.Z.; data curation: J.A.W. and D.S.; writing—original draft preparation: J.A.W., J.F.-F., P.M., and J.P.; writing—review and editing: J.A.W.; project administration, J.P. and R.D. All authors have read and agreed to the published version of the manuscript.

Funding: This research was funded by the Polish National Centre for Research and Development (*Narodowe Centrum Badań i Rozwoju*), grant number POIR.04.01.02-00-0008/18.

Institutional Review Board Statement: All animal procedures were approved by the Local Ethical Committee for Animal Research (Poznań, Poland; decision no. 57/2020).

Informed Consent Statement: Not applicable.

Data Availability Statement: The data presented in this study are available on request from the corresponding author.

Conflicts of Interest: The authors declare that they have no conflicts of interest. The funders had no role in the design of the study; in the collection, analysis, or interpretation of data; in the writing of the manuscript; or in the decision to publish the results.

References

1. Katsoulos, P.D.; Karatzia, M.A.; Dovas, C.I.; Filioussis, G.; Papadopoulos, E.; Kiossis, E.; Arsenopoulos, K.; Papadopoulos, T.; Boscos, C.; Karatzias, H. Evaluation of the in-field efficacy of oregano essential oil administration on the control of neonatal diarrhea syndrome in calves. *Res. Vet. Sci.* **2017**, *115*, 478–483. [CrossRef] [PubMed]
2. Bampidis, V.A.; Christodoulou, V.; Florou-Paneri, P.; Christaki, E. Effect of dried oregano leaves versus neomycin in treating newborn calves with colibacillosis. *J. Vet. Med. A Physiol. Pathol. Clin. Med.* **2006**, *53*, 154–156. [CrossRef] [PubMed]
3. Stefańska, B.; Sroka, J.; Katzer, F.; Goliński, P.; Nowak, W. The effect of probiotics, phytobiotics and their combination as feed additives in the diet of dairy calves on performance, rumen fermentation and blood metabolites during the preweaning period. *Anim. Feed Sci. Tech.* **2021**, *272*, 114738. [CrossRef]
4. Grela, E.R.; Klebaniuk, R.; Kwiecień, M.; Pietrzak, K. Phytobiotics in Animal Production. *Przegląd Hod.* **2013**, *3*, 21–24. (In Polish)
5. Kurzeja, E.; Stec, M.; Kiryk, M.; Maly, B.; Misiek, K.; Sołujan, A. Changes in the antioxidant properties of herbs under the influence of steam sterilization and storage. *Bromat. Chem. Toksykol.* **2012**, *3*, 980–984. (In Polish)
6. Kalisz, S.; Ścibisz, I. The effect of the addition of plant extracts on the content of total polyphenols, anthocyanins, vitamin C and the antioxidant capacity of blackcurrant nectars. *Żywność. Nauka. Technol. Jakość* **2010**, *5*, 45–55. (In Polish)
7. Szczucińska, A.; Kurzepa, K.; Kleczkowska, P.; Lipkowski, A.W. Technological aspects of milk thistle endosperm for use as antioxidant additives. *Rośliny Oleiste* **2006**, *27*, 357–366. (In Polish)
8. Dragland, S.; Senoo, H.; Wake, K.; Holte, K.; Blomhoff, R. Several culinary and medicinal herbs are important sources of dietary antioxidants. *J. Nutr.* **2003**, *133*, 1286–1290. [CrossRef] [PubMed]
9. Wójtowski, J.; Danków, R.; Foksowicz-Flaczyk, J.; Grajek, K. Herbal additives in the nutrition of cows, sheep and dairy goats. *Życie Weter.* **2019**, *94*, 550–556. (In Polish)
10. Mastellone, V.; Morittu, V.M.; Musco, N.; Spina, A.A.; Malgeri, A.; Molinari, M.L.; D'Aniello, B.; Infascelli, F.; Tudisco, R.; Lombardi, P. Dietary supplementation with a phytocomplex affects blood parameters and milk yield and quality in grazing goats. *Small Rumin. Res.* **2021**, *201*, 106421. [CrossRef]
11. Simitzis, P.E.; Feggeros, K.; Bizelis, J.A.; Deligeorgis, S.C. Behavioral reaction to essential oils supplementation in sheep. *Biotech. Anim. Husb.* **2005**, *5–6*, 91–103. [CrossRef]
12. Chen, S.; Luo, S.; Yan, C. Gut Microbiota Implications for Health and Welfare in Farm Animals: A Review. *Animals* **2022**, *12*, 93. [CrossRef] [PubMed]

13. National Center for Biotechnology Information. Available online: https://www.ncbi.nlm.nih.gov/Taxonomy/Browser/wwwtax.cgi?id=1578 (accessed on 25 October 2017).
14. Fontana, L.; Bermudez-Brito, M.; Plaza-Diaz, J.; Munoz-Quezada, S.; Gil, A. Sources, isolation, characterisation and evaluation of probiotics. *Br. J. Nutr.* **2013**, *109* (Suppl. 2), S35–S50. [CrossRef]
15. Jarzynowska, A.; Peter, E. The effect of adding herbs to the summer diet on the fatty acid profile of the lipid fraction of sheep milk. *Rocz. Nauk. Pol. Tow. Zootech.* **2017**, *13*, 31–42. (In Polish) [CrossRef]
16. Kowalski, Z.M. Goat feeding. In *Breeding, Housing and Use of Goats*, 3rd ed.; Wójtowski, J.A., Ed.; Publishing House of the University of Life Sciences in Poznań: Poznań, Poland, 2021; (In Polish). ISBN 978-83-7160-985-5.
17. Isolauri, E.; Salminen, S.; Ouwehand, A.C. Microbial-gut interactions in health and disease. Probiotics. *Best Pract. Res. Clin. Gastroenterol.* **2004**, *18*, 299–313. [CrossRef]
18. de Vos, W.M. Systems solutions by lactic acid bacteria: From paradigms to practice. *Microb. Cell Fact.* **2011**, *10* (Suppl. 1), S2. [CrossRef]
19. Song, Y.; Malmuthuge, N.; Steele, M.A.; Guan, L.L. Shift of hindgut microbiota and microbial short chain fatty acids profiles in dairy calves from birth to pre-weaning. *FEMS Microbiol. Ecol.* **2018**, *94*, 1–15. [CrossRef]
20. Dill-McFarland, K.A.; Weimer, P.J.; Breaker, J.D.; Suen, G. Diet influences early microbiota development in dairy calves without long-term impacts on milk production. *Appl. Environ. Microbiol.* **2019**, *85*, e02141. [CrossRef]
21. Wang, B.; Ma, M.P.; Diao, Q.Y.; Tu, Y. Saponin-induced shifts in the rumen microbiome and metabolome of young cattle. *Front. Microbiol.* **2019**, *10*, 356. [CrossRef]
22. Behr, C.; Sperber, S.; Jiang, X.; Strauss, V.; Kamp, H.; Walk, T.; Herold, M.; Beekmann, K.; Rietjens, I.; van Ravenzwaay, B. Microbiome-related metabolite changes in gut tissue, cecum content and feces of rats treated with antibiotics. *Toxicol. Appl. Pharmacol.* **2018**, *355*, 198–210. [CrossRef] [PubMed]
23. Oultram, J.; Phipps, E.; Teixeira, A.G.; Foditsch, C.; Bicalho, M.L.; Machado, V.S.; Bicalho, R.C.; Oikonomou, G. Effects of antibiotics (oxytetracycline, florfenicol or tulathromycin) on neonatal calves' faecal microbial diversity. *Vet. Rec.* **2015**, *177*, 598. [CrossRef]
24. Yousif, M.H.; Li, J.H.; Li, Z.Q.; Maswayi Alugongo, G.; Ji, S.K.; Li, Y.X.; Wang, Y.J.; Li, S.L.; Cao, Z.J. Low concentration of antibiotics modulates gut microbiota at different levels in pre-weaning dairy calves. *Microorganisms* **2018**, *6*, 118. [CrossRef]
25. Gomez, D.E.; Arroyo, L.G.; Costa, M.C.; Viel, L.; Weese, J.S. Characterization of the fecal bacterial microbiota of healthy and diarrheic dairy calves. *J. Vet. Int. Med.* **2017**, *31*, 928–939. [CrossRef]
26. Zhang, J.; Shi, H.T.; Wang, Y.J.; Cao, Z.J.; Yang, H.J.; Li, S.L. Effect of limit-fed diets with different forage to concentrate ratios on fecal bacterial and archaeal community composition in holstein heifers. *Front. Microbiol.* **2018**, *9*, 976. [CrossRef]
27. Kim, M.; Kim, J.; Kuehn, L.A.; Bono, J.L.; Berry, E.D.; Kalchayanand, N.; Freetly, H.C.; Benson, A.K.; Wells, J.E. Investigation of bacterial diversity in the feces of cattle fed different diets. *J. Anim. Sci.* **2014**, *92*, 683–694. [CrossRef] [PubMed]
28. Huang, S.; Ji, S.; Wang, F.; Huang, J.; Alugongo, G.M.; Li, S. Dynamic changes of the fecal bacterial community in dairy cows during early lactation. *AMB Expr.* **2020**, *10*, 167. [CrossRef]
29. Draksler, D.; Locascio, M.; González, S.; Oliver, G. The development of faecal flora in young Creole goats. *Small Rumin. Res.* **2002**, *46*, 67–70. [CrossRef]
30. Stella, A.V.; Paratte, R.; Valnegri, L.; Cigalino, G.; Soncini, G.; Chevaux, E.; Dell'Orto, V.; Savoini, G. Effect of administration of live Saccharomyces cerevisiae on milk production, milk composition, blood metabolites, and faecal flora in early lactating dairy goats. *Small Rumin. Res.* **2007**, *67*, 7–13. [CrossRef]
31. Shokryazdan, P.; Sieo, C.C.; Kalavathy, R.; Boo Liang, J.; Banu Alitheen, N.; Jahromi, M.F.; Yin Ho, Y. Probiotic potential of *Lactobacillus* strains with antimicrobial activity against some human pathogenic strains. *BioMed Res. Int.* **2014**, *2014*, 927268. [CrossRef]
32. Jiao, J.; Wu, J.; Zhou, C.; Tang, S.; Wang, M.; Tan, Z. Composition of ileal bacterial community in grazing goats varies across non-rumination, transition and rumination stages of life. *Front. Microbiol.* **2016**, *7*, 1364. [CrossRef] [PubMed]

animals

MDPI

Article

The Effects of Fucoidan Dietary Supplementation on Growth Performance, Serum Antioxidant Capacity, Immune Function Indices and Intestinal Morphology in Weaned Kids

Weiguang Yang †, Jiayi Chen †, Guangzhen Guo, Shengnan Wang, Su Peng, Zhenhua Gao, Zhihui Zhao, Ruixia Lan * and Fuquan Yin *

College of Coastal Agriculture Science, Guangdong Ocean University, Zhanjiang 524088, China; yangweiguang1@stu.gdou.edu.cn (W.Y.); chen17852729978@163.com (J.C.); guoguangzhen1@stu.gdou.edu.cn (G.G.); 2112004098@stu.gdou.edu.cn (S.W.); sugar7364@126.com (S.P.); xmsgzhh@126.com (Z.G.); zhzhao@gdou.stu.cn (Z.Z.)

* Correspondence: lanrx@gdou.edu.cn (R.L.); yinfuquan01@163.com (F.Y.)

† These authors contributed equally to this work.

Simple Summary: During the weaning period, the change of feed and separation from ewe induce weaning stress and may affect the growth and health of kids. The application of antibiotics could relieve weaning stress; however, their prophylactic application coerces researchers to find antibiotic alternatives to relieve weaning stress. Fucoidan is a natural plant extract widely used in animal production with antioxidant and immune-modulatory properties resulting in beneficial effects on the intestinal tract. In the present study, fucoidan dietary supplementation boosted antioxidant and immune functions, improved the morphology of the intestinal tract and promoted the growth performance of kids. These results indicated that fucoidan could be used to alleviate weaning stress in kids.

Abstract: The purpose of this study was to evaluate the effects of fucoidan dietary supplementation on growth performance, organs' relative weight, serum anti-oxidation markers, immune function indices and intestinal morphology in weaned kids. A total of 60 2-month-old weaned castrated male kids (Chuanzhong black goat) were used for this 30-day experiment and randomly allocated to four groups. The control group (CON) fed a basal diet, while the other three groups were provided with the same diet further supplemented with fucoidan at 0.1%, 0.3% or 0.5%, namely, F1, F2 and F3 groups, respectively. The results indicated that dietary fucoidan supplementation significantly increased ($p < 0.05$) the activity of catalase (CAT) when compared to the CON group on day 15. Moreover, the addition of fucoidan at 0.3% and 0.5% significantly increased ($p < 0.05$) the activities of glutathione peroxidase (GSH-Px) and total superoxide dismutase (T-SOD). On day 30, dietary fucoidan supplementation significantly reduced ($p < 0.05$) the feed conversion rate (FCR), contents of tumor necrosis-α (TNF-α), interleukin-1β (IL-1β) and interleukin-6 (IL-6), while it significantly increased ($p < 0.05$) the activity of total superoxide dismutase (T-SOD), the content of immunoglobulin G (IgG) and the villus height (VH) of the duodenum. Moreover, dietary 0.3% and 0.5% fucoidan supplementation significantly increased ($p < 0.05$) the villus height (VH) of the jejunum and ileum and significantly reduced ($p < 0.05$) the crypt depth (CD) of ileum. In conclusion, dietary fucoidan had positive effects on growth performance, serum anti-oxidation, immune function and intestinal morphology of weaned kids.

Keywords: fucoidan; weaned kids; growth performance; antioxidant capacity; immune function

Citation: Yang, W.; Chen, J.; Guo, G.; Wang, S.; Peng, S.; Gao, Z.; Zhao, Z.; Lan, R.; Yin, F. The Effects of Fucoidan Dietary Supplementation on Growth Performance, Serum Antioxidant Capacity, Immune Function Indices and Intestinal Morphology in Weaned Kids. *Animals* **2022**, *12*, 574. https://doi.org/10.3390/ani12050574

Academic Editors: Panagiotis Simitzis and Athanasios I. Gelasakis

Received: 23 December 2021
Accepted: 21 February 2022
Published: 24 February 2022

1. Introduction

In modern small ruminants' production, weaning stress is an inevitable problem observed in small animals. Weaning stress had negative effects on antioxidant capacity, immunity, intestinal morphology and growth performance [1–3]. Formerly, antibiotics were widely used to alleviate weaning stress [4]. However, the ban of antibiotic use in feed worldwide forces researchers to find antibiotic alternatives. Fucoidan, a kind of macromolecular polysaccharide rich in sulfate, occurs in the cell walls of brown algae, mucous matrix and some marine invertebrates [5,6]. Fucoidan has been proven to possess many biological properties, including immunomodulatory, antioxidant and antibacterial properties, resulting in the promotion of animal growth [7–10]. Thus, fucoidan was widely used in functional foods and animal production. Dietary fucoidan administration increased feed intake, daily gain and feed efficiency and also had beneficial effects on intestinal morphology, antioxidant capacity and immune function in weaning pigs [11–13]. Similar results are also reported in chickens and fish [14–16]. In summary, fucoidan could be used as an environmentally friendly substitute for antibiotics in diets to improve growth in pigs, chicken and fish. However, data on the application of fucoidan in small ruminants are scarce. Therefore, the purpose of this study was to evaluate the effects of fucoidan on growth performance, organs' relative weight, antioxidant markers, immunity indices and intestine morphology in weaned kids.

2. Materials and Methods

2.1. Animals, Diet and Experimental Design

The experimental protocol applied in this study followed the guidelines of the Animal Care and Use Committee of Guangdong Ocean University.

The fucoidan used in this study was provided by a company (Mingyue Hailin Fucoidan Biotechnology Co., Ltd., Qingdao, China). Fucoidan had the form of a powder with a yellow color and a smell of seaweed. The purity of fucoidan was 98%, and the sulfate ion content was 28.9%.

A total of 60 two-month-old weaned castrated male kids (Chuanzhong black goat) with an average initial body weight of 12.5 $\pm$ 0.5 kg were used in this 30-day experiment. Kids were weaned at 60 days and randomly allocated to 4 treatments with 15 replications. The control group (CON) was fed with a basal diet, while the other three treatment groups were fed with the same diet further supplemented with fucoidan at 0.1% (F1 group), 0.3% (F2 group) and 0.5% (F3 group). Three kids were placed in a pen (each pen = 3.1 m $\times$ 2.5 m $\times$ 1 m). The basal diet (Table 1) was formulated to meet or exceed the nutrient requirement of the Feeding standard of Goat, China (NY/T 861-2004).

Table 1. The composition and level of basal diet (air-dry basis).

Items	Content
Ingredients (%)	
Pennisetum purpureum	35.00
Corn	40.89
Soybean meal	13.98
Wheat bran	7.15
NaCl	0.65
$CaHPO_4$	0.84
Limestone	0.84
Premix [1]	0.65
Total	100
Nutrient level	
DM (%)	88.86
ME [2] (MJ/Kg)	10.43
CP (%)	12.06
NDF (%)	30.06
ADF (%)	16.39
Ca (%)	0.76
P (%)	0.54

[1] The premix provided the following per kg of diets: VA 8 000 IU, VD 2 000 IU, VE 40 IU, Cu 12 mg, Fe 70 mg, Mn 50 mg, Zn 80 mg, I 1.0 mg, Se 0.27 mg and Co 0.3 mg. [2] ME was a calculated value.

Before the trial started, the sheep house was cleaned and sterilized. A 7-day pre-trial was conducted first, during which vaccination, deworming and numbering were performed. The roughage consisted of silage and *Aneurolepidium* Chinese hay. The kids were fed twice a day at 8:30 am and 17:30 pm, with access to clean drinking water available ad libitum. Fucoidan was manually mixed into the concentrate. The kids were fed concentrate first and then roughage.

2.2. Growth Performance

Body weight was determined on day 1, 15 and 30, the feed intake was recorded daily. The average daily gain (ADG), average daily feed intake (ADFI) and feed conversion rate (FCR) were also calculated.

2.3. Sample Collection and Organs' Relative Weight

On days 15 and 30, blood samples were collected from the jugular vein, and then they were centrifuged at 3500 g for 10 min (4 °C). The serum was collected and stored at −20 °C for later analysis.

Kids were fasted for 12 h prior to slaughter at the end of the trial. Six kids of similar weight per group were selected for slaughter. About 2.5 cm segment of the duodenum, jejunum and ileum were trimmed and used for morphological indices. Samples of tissue were washed with PBS, then fixed in paraformaldehyde for histological evaluations. Finally, organs were weighed.

The organ relative weight was calculated by the following formula: organ index (%) = organ weight/body weight × 100%.

2.4. Serum Antioxidant

The activities of total antioxidant capacity (T-AOC), total superoxide dismutase (T-SOD), catalase (CAT), glutathione peroxidase (GSH-Px) and malondialdehyde (MDA) content were measured using commercial kits according to manufacturer's guidelines (Nanjing Jiancheng Bioengineering Institute, Jiangsu, China).

2.5. Serum Immunity

The contents of immunoglobulin G (IgG), interleukin-1β (IL-1β), interleukin-6 (IL-6), interleukin-2 (IL-2), interleukin-10 (IL-10) and tumor necrosis factor-α (TNF-α) were estimated by enzyme-linked immunosorbent assay kits according to manufacturer's guidelines (Jiangsu Meimian industrial Co., Ltd., Jiangsu, China).

2.6. Intestinal Histomorphology

The samples from the duodenum, jejunum and ileum were fixed in paraformaldehyde for 24 h at room temperature and subsequently dehydrated through a graded ethanol series, cleared with xylene and embedded in paraffin. Then, tissues were cut into 5 μm-thick continuous sections. Finally, the sections were stained with hematoxylin for 2 min and eosin for 40 s, and then dehydrated and mounted on slides. The morphological parameters were measured by Image Pro Plus 6.0 software (Media Cybernetics, Silver Spring, MD). The morphological parameters of the intestinal tract included villus height (VH), crypt depth (CD) and radio of villus height to crypt depth (VCR).

2.7. Statistical Analysis

All statistical analyses were performed using SPSS 26.0 via one-way ANOVA, and differences were detected by Ducan's multi-range test. The results are expressed as mean ± standard error of the mean (SEM), and differences are considered significant at $p < 0.05$.

3. Results

3.1. Grouth Performance

As shown in Table 2, kids fed F2 and F3 diets had a higher final body weight than those fed the CON and F1 diets. Kids fed diets with fucoidan significantly reduced ($p < 0.05$) FCR compared to those fed the CON diet during days 16 to 30 and the overall period. Kids fed F2 and F3 diets had significantly higher ($p < 0.05$) values for ADG and ADFI than those fed the CON diet during days 16 to 30 and the overall period. No significant differences were observed for growth performance among treatments during days 1 to 15.

Table 2. Effects of fucoidan administration on growth performance in weaned kids.

Item [1]	CON	F1	F2	F3	SEM [2]	*p*-Value
BW of day 1 (kg)	12.20	11.38	12.27	12.98	0.99	0.26
BW of day 15 (kg)	12.47	11.53	13.25	12.78	1.08	0.12
BW of day 30 (kg)	12.85 [b]	12.88 [b]	15.28 [a]	15.13 [a]	1.09	0.02
Day 1 to 15						
ADG (g/d)	30.00	31.78	51.78	53.22	10.41	0.10
ADFI (g/d)	432.01	427.02	484.05	445.01	24.41	0.16
FCR (g)	14.52	12.01	9.41	11.62	1.61	0.08
Day 16 to 30						
ADG (g/d)	80.01 [a]	110.01 [a]	141.78 [b]	121.04 [b]	13.90	0.01
ADFI (g/d)	569.05 [a]	620.67 [b]	711.03 [c]	863.04 [d]	16.08	<0.01
FCR (g)	9.97 [b]	5.79 [a]	5.02 [a]	6.22 [a]	0.69	<0.01
Overall						
ADG (g/d)	45.00 [a]	50.06 [a]	100.33 [c]	72.01 [b]	9.71	0.02
ADFI (g/d)	501.09 [a]	525.08 [a]	597.01 [b]	565.02 [b]	22.71	0.01
FCR (g)	11.25 [c]	10.60 [b]	6.05 [a]	8.25 [a]	1.08	0.01

[1] CON, basal diet; F1, basal diet +0.1% fucoidan; F2, basal diet +0.3% fucoidan; F3, basal diet +0.5% fucoidan. BW, body weight; ADG, average daily gain; ADFI, average daily feed intake; FCR, feed conversion rate. [2] SEM, standard error of the mean. [a–d] Values in the same row with different letters are significantly different ($p < 0.05$). Results are presented as mean ± SEM ($n = 6$).

3.2. Organs' Relative Weight

As shown in Table 3, no significant differences were observed for organs' relative weight among the experimental groups.

Table 3. Effects of fucoidan administration on organ index in weaned kids.

Item [1], %	CON	F1	F2	F3	SEM [2]	*p*-Value
Heart index	0.41	0.40	0.39	0.40	0.50	0.13
Liver index	1.56	1.85	1.52	1.65	0.14	0.15
Spleen index	0.14	0.15	0.15	0.13	0.01	0.22
Lung index	1.15	1.37	1.170	1.37	0.20	0.61
Kidney index	0.33	0.35	0.33	0.33	0.03	0.91
Small intestine index	4.32	4.10	4.21	4.21	0.30	0.06

[1] CON, basal diet; F1, basal diet +0.1% fucoidan; F2, basal diet +0.3% fucoidan; F3, basal diet +0.5% fucoidan. [2] SEM, standard error of mean.

3.3. Serum Antioxidant Capacity

As shown in Figure 1, on day 15, kids fed diets with fucoidan had a significantly increased CAT activity ($p < 0.05$). Lambs fed F2 and F3 diets showed significantly increased ($p < 0.05$) activity of GSH-Px and T-SOD than those kids fed CON and F1 diets. No significant differences were observed for MDA and T-AOC among treatments.

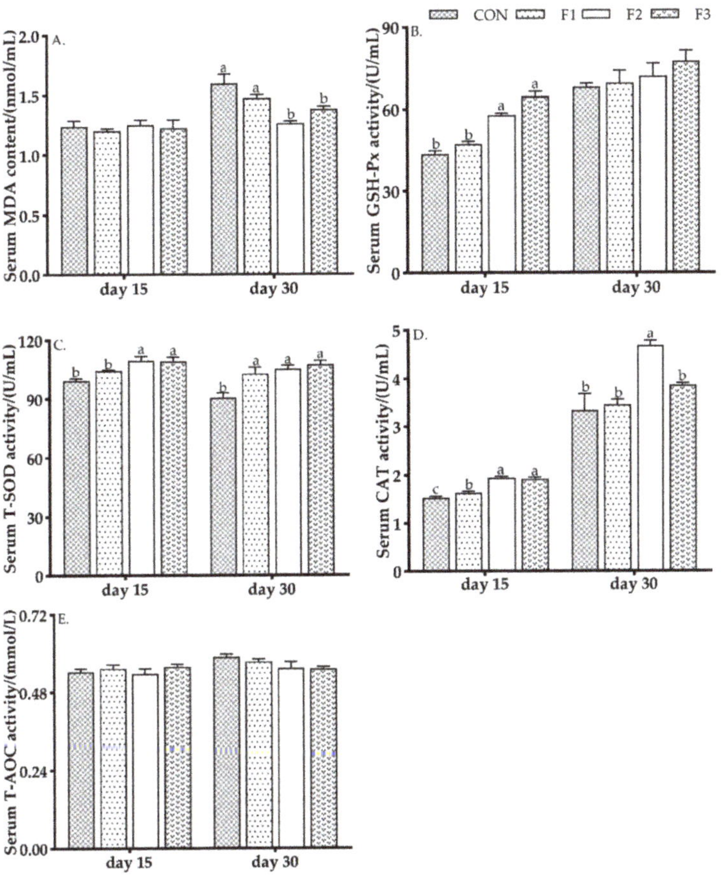

Figure 1. Effects of dietary fucoidan administration on antioxidant indexes in weaned kids. (**A**) serum MDA content; (**B**) serum GSH-Px activity; (**C**) serum T-SOD activity; (**D**) serum CAT activity; (**E**) serum T-AOC activity. Results are presented as mean ± SEM (n = 6). $^{a, b, c}$ means significantly different ($p < 0.05$). CON, basal diet; F1, basal diet +0.1% fucoidan; F2, basal diet +0.3% fucoidan; F3, basal diet +0.5% fucoidan. MDA, malondialdehyde; GSH-Px, glutathione peroxidase; T-SOD, total superoxide dismutase; CAT, catalase; T-AOC, total antioxidant capacity.

On day 30, kids fed diets with fucoidan showed significantly increased ($p < 0.05$) activity of T-SOD. Kids fed F2 and F3 diets had a significantly reduced ($p < 0.05$) MDA content than those fed CON and F1 diets. Kids fed F2 diet had a significantly increased ($p < 0.05$) activity of CAT than those kids fed CON, F1 and F2 diets. No significant differences were observed for GSH-Px and T-AOC content among treatments.

3.4. Serum Immuntiy Indices

As shown in Figure 2, kids fed diets with fucoidan had significantly higher IgG content than that fed with CON diet on day 30 ($p < 0.05$). Kids fed diets with fucoidan had significantly lower TNF-α, IL-1β and IL-6 contents than those fed CON diet ($p < 0.05$). On the other hand, F2 and F3 animals had significantly higher IL-2 and IL-10 contents than those fed CON and F1 diets ($p < 0.05$).

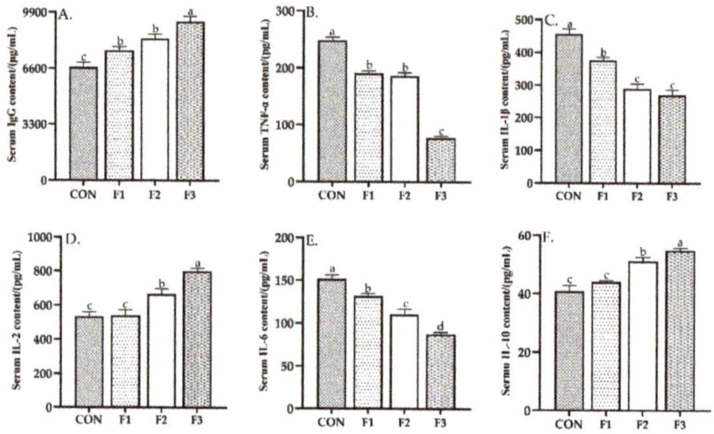

Figure 2. Effects of dietary fucoidan administration for 30 days on serum immunity indexes in weaned kids. (**A**) serum IgG content; (**B**) serum TNF-α content; (**C**) serum IL-1β content; (**D**) serum IL-2 content; (**E**) serum IL-6 content; (**F**) serum IL-10 content. Results are presented as mean ± SEM (n = 6). $^{a–d}$ means significantly different ($p < 0.05$). CON, basal diet; F1, basal diet +0.1% fucoidan; F2, basal diet +0.3% fucoidan; F3, basal diet +0.5% fucoidan. IgG, Immunoglobulin G; TNF-α, tumor necrosis factor-α; IL-1β, interleukin-1β; IL-2, interleukin-2; IL-6, interleukin-6; IL-10, interleukin-10.

3.5. *Intestinal Morphology*

The effects of dietary fucoidan administration on intestinal morphology are illustrated in Figure 3. From the HE staining, we could see that the intestinal villi were denser and longer than the CON in the duodenum, jejunum and ileum. As shown in Table 4, kids fed with fucoidan-supplemented diets had significantly higher VH in the duodenum on day 30 ($p < 0.05$). Moreover, kids fed with F2 and F3 diets significantly increased ($p < 0.05$) the VCR in the duodenum and ileum than those fed with CON and F1 diets. Kids fed with F2 and F3 diets had significantly lower ($p < 0.05$) CD values in the ileum than those fed with CON and F1 diets. Kids fed with F2 and F3 diets significantly increased ($p < 0.05$) the VH in the ileum than those fed with CON and F1 diets. However, no significant differences were observed for CD in the duodenum and CD or VCR in the jejunum among treatments.

Table 4. Effects of dietary fucoidan administration on intestinal morphology in weaned kids.

Item [1]	CON	F1	F2	F3	SEM [2]	p-Value
Duodenum						
VH (μm)	514.83 ^{c}	657.89 ^{b}	841.39 ^{a}	904.28 ^{a}	48.38	<0.01
CD (μm)	420.44	358.94	407.33	244.56	68.56	0.07
VCR	1.42 ^{b}	1.94 ^{b}	2.21 ^{b}	3.02 ^{a}	0.51	<0.01
Jejunum						
VH (μm)	578.50 ^{b}	657.06 ^{b}	708.28 ^{a}	774.39 ^{a}	43.41	<0.01
CD (μm)	345.11	337.67	412.45	467.00	65.75	0.19
VCR	1.70	2.15	1.94	1.75	0.36	0.59
Ileum						
VH (μm)	578.50 ^{b}	657.06 ^{b}	708.28 ^{a}	771.50 ^{a}	44.05	<0.01
CD (μm)	420.45 ^{a}	435.61 ^{a}	314.22 ^{b}	354.00 ^{b}	85.80	0.04
VCR	1.87 ^{b}	1.76 ^{b}	2.44 ^{b}	3.82 ^{a}	0.58	0.02

[1] CON, basal diet; F1, basal diet +0.1% fucoidan; F2, basal diet +0.3% fucoidan; F3, basal diet +0.5% fucoidan. VH, villus height; CD, crypt depth; VCR, ratio of villus height to crypt depth. [2] SEM, standard error of mean. $^{a–c}$ Values in the same row with different letters are significantly different ($p < 0.05$). Results are presented as mean ± SEM (n = 6).

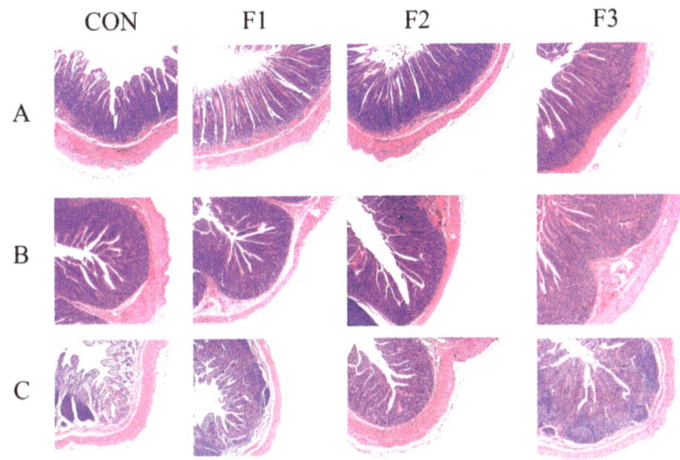

Figure 3. Micrograph of duodenum, jejunum and ileum supplemented with fucoidan in the diet of weaned kids: (**A**) duodenum; (**B**) jejunum; (**C**) ileum. F1, basal diet +0.1% fucoidan; F2, basal +0.3% fucoidan; F3, basal +0.5% fucoidan.

4. Discussion

4.1. Growth Performance

In general, weaning stress adversely affects the growth performance of kids by reducing feed intake, feed efficiency, immune suppression and increasing intestinal damage [17,18]. In our study, dietary fucoidan administration had no significant effects on growth performance among treatments during days 1 to 15. In support of our results, Rattigan et al. [19] reported that dietary fucoidan supplementation had no significant effects on feed intake and daily gain during days 1 to 14; however, supplemented weaner pigs showed higher feed intake and ADG than those in the CON group during days 16 to 30 and 1 to 30. Similarly, Draper et al. [20] indicated that dietary fucoidan administration increased feed intake and decreased FCR in weaning pigs. We speculated that the improved effects of fucoidan on the growth performance in this study might be related to the improvement of feed intake, antioxidant capacity, immune function and intestinal structure.

4.2. Serum Antioxidant

Weaning can lead to the excessive production of oxygen-free radicals, which results in oxidative stress and is linked with reduced growth performance [21,22]. SOD, GSH-Px and CAT activities are the first lines against oxidative injury, and MDA was the final product of lipid peroxidation [23,24]. In this study, dietary fucoidan administration increased the activities of GSH-Px, SOD and CAT on day 15 and decreased the content of MDA on day 30. Similarly, Yang et al. [25] reported that dietary supplementation with 0.1% fucoidan increased the activities of CAT and SOD and decreased the content of MDA in *Pelteobagrus fulvidraco*. Zhang et al. [26] indicated that supplementation with fucoidan improved the activities of SOD, CAT and GSH and reduced the content of MDA in drosophila geriatric. In summary, the improvement of antioxidant capacity with dietary fucoidan supplementation might be related to the increased activities of antioxidant enzymes and the reduced lipid oxidation values.

4.3. Serum Immunity

Previous research confirmed that weaning reduced immunity function, making invasion by external pathogenic microorganisms easier [27]. Immunoglobulin and cytokines are an important part of the immune system [28]. The immunoglobulins protect the body by removing pathogenic microorganisms and harmful molecules [29]. Cytokines play a key role in inflammation and anti-inflammatory processes [30]. In this study, dietary fucoidan administration increased the content of IgG, IL-2 and IL-10 in serum while reducing the content of IL-1β, TNF-α and IL-6. In agreement with our study, Lean et al. [31] reported that dietary fucoidan administration reduced the content of IL-1β in colons in mice with colitis. Tomori et al. [32] indicated that fucoidan increased the contents of IgG and IL-2 in serum by promoting the proliferation of immune cells and reducing the contents of IL-4 and IL-5 in spleens of mice. Moreover, Delma et al. [33] indicated that fucoidan exerted anti-inflammatory activity by regulating the NF-κB signaling pathway and reducing the expression of p53. Taken together, the improvement of immunological functions with dietary fucoidan supplementation may be related to the regulation of inflammatory-related factors release.

4.4. Intestinal Morphology

At weaning, the reduced villus height and increased crypt depth are related to stress [34,35]. In this study, dietary fucoidan administration increased the villus height of the duodenum, jejunum and ileum and decreased the crypt depth of the ileum. Our results are in line with that of Leonard et al. [36], who reported that dietary fucoidan administration in weaning pigs increases villus height and VCR of jejunum. Similarly, Walsh et al. [12] indicated that dietary fucoidan supplementation increased the villus height of small intestinal in post-weaning pigs. According to current studies, it indicated that dietary fucoidan administration could improve the intestinal morphology of the small intestine.

5. Conclusions

In summary, this study showed that fucoidan dietary supplementation improved the feed intake, daily gain, antioxidant capacity, immune status and intestinal morphology in weaned kids.

Author Contributions: Conceptualization, F.Y. and W.Y.; methodology, F.Y., R.L. and W.Y.; software, J.C., S.W. and S.P.; validation, F.Y. and Z.Z.; formal analysis, W.Y.; investigation, G.G.; resources, F.Y. and Z.G.; data curation, W.Y.; writing—original draft preparation, W.Y.; writing—review and editing, F.Y.; visualization, W.Y.; project administration, F.Y.; funding acquisition, F.Y. All authors have read and agreed to the published version of the manuscript.

Funding: This research was funded by the Fundamental Research Funds for the Guangdong Province (2020A03025), the Fundamental Research Funds for the Agriculture and Rural Affairs of Guangdong Department (2019KJ127), the Fundamental Research Funds for Education Department of Guangdong (2021003).

Institutional Review Board Statement: This research was approved by the Animal Care and Use Committee of Guangdong Ocean University (SYXK-2018-0147, 2018).

Data Availability Statement: The data presented in this study are available on request from the corresponding author.

Acknowledgments: We would like to thank all people who were involved in the trial. We are grateful for the financial support from the Agriculture and Rural Affairs of Guangdong Department.

Conflicts of Interest: The authors declare no conflict of interest.

References

1. Han, Y.S.; Tang, C.H.; Li, Y.; Yu, Y.A.; Zhan, T.F.; Zhao, Q.Y.; Zhang, J.M. Effects of Dietary Supplementation with Clostridium butyricum on Growth Performance, Serum Immunity, Intestinal Morphology, and Microbiota as an Antibiotic Alternative in Weaned Piglets. *Animals* **2020**, *10*, 2287. [CrossRef] [PubMed]
2. Zhang, M.Y.; Hou, G.J.; Hu, P.; Feng, D.; Wang, J.; Zhu, W.Y. Nano chitosan-zinc complex improves the growth performance and antioxidant capacity of the small intestine in weaned piglets. *Br. J. Nutr.* **2020**, *126*, 31–35. [CrossRef] [PubMed]
3. Gabler, N.K.; Helm, E.T.; Mille, C.D. Impact of weaning stress, disease, and diet on pig performance, intestinal function and integrity. *J. Anim. Sci.* **2019**, *97*, 31–32. [CrossRef]
4. Lourenco, J.M.; Hampton, R.S.; Johnson, H.M.; Callaway, T.R.; Rothrock, M.J.; Azain, M.J. The Effects of Feeding Antibiotic on the Intestinal Microbiota of Weanling Pigs. *Front. Vet. Sci.* **2021**, *8*, 131. [CrossRef]

5. Liu, X.; Xi, X.Y.; Jia, A.R.; Zhang, M.S.; Cui, T.T.; Bai, X.F.; Shi, Y.P.; Liu, C.H. A fucoidan from Sargassum fusiforme with novel structure and its regulatory effects on intestinal microbiota in high-fat diet-fed mice. *Food Chem.* **2021**, *358*, 129908. [CrossRef]
6. Zhu, Y.L.; Liu, L.B.; Sun, Z.Y.; Ji, Y.J.; Wang, D.Y.; Mei, L.; Shen, P.L.; Li, Z.X.; Tang, S.; Zhang, H.; et al. Fucoidan as a marine-origin prebiotic modulates the growth and antibacterial ability of Lactobacillus rhamnosus. *Int. J. Biol. Macromol.* **2021**, *180*, 599–607. [CrossRef]
7. Hou, Y.; Wang, J.; Jin, W.H.; Zhang, H.; Zhang, Q.B. Degradation of Laminaria japonica fucoidan by hydrogen peroxide and antioxidant activities of the degradation products of different molecular weights. *Carbohydr. Polym.* **2012**, *87*, 153–159. [CrossRef]
8. Gora, A.H.; Sahu, N.P.; Sahoo, S.; Rehman, S.; Dar, S.A.; Ahmad, I.; Agarwal, D. Effect of dietary Sargassum wightii and its fucoidan-rich extract on growth, immunity, disease resistance and antimicrobial peptide gene expression in Labeo rohita. *Int. Aquat. Res.* **2018**, *10*, 115–131. [CrossRef]
9. Traifalgar, R.F.; Serrano, A.E.; Corre, V.; Kira, H.; Tung, H.T.; Michael, F.R.; Kader, M.A.; Laining, A.; Yokoyama, S.; Ishikawa, M.; et al. Evaluation of Dietary Fucoidan Supplementation Effects on Growth Performance and Vibriosis Resistance of Penaeus monodon Postlarvae. *Aquac. Sci.* **2009**, *57*, 167–174.
10. Shokaiyan, M.; Ashayerizadeh, O.; Shams, S.M.; Dastar, B. Algal Crude Fucoidan Alone or with Bacillus subtilisDSM 17299in Broiler Chickens Diet: Growth Performance, Carcass Characteristics, Blood Metabolites, and Morphology of Intestine. *Poult. Sci. J.* **2019**, *7*, 87–94.
11. Reilly, P.; O'Doherty, J.V.; Pierce, K.M.; Callan, J.J.; O'Sullivan, J.T.; Sweeney, T. The effects of seaweed extract inclusion on gut morphology, selected intestinal microbiota, nutrient digestibility, volatile fatty acid concentrations and the immune status of the weaned pig. *Animal* **2008**, *2*, 1465–1473. [CrossRef]
12. Walsh, A.M.; Sweeney, T.; O'Shea, C.J.; Doyle, D.N.; O'Doherty, J.V. Effects of supplementing dietary laminarin and fucoidan on intestinal morphology and the immune gene expression in the weaned pig. *J. Anim. Sci.* **2012**, *90*, 284–286. [CrossRef] [PubMed]
13. McDonnell, P.; Figat, S.; O'Doherty, J.V. The effect of dietary laminarin and fucoidan in the diet of the weanling piglet on performance, selected faecal microbial populations and volatile fatty acid concentrations. *Animal* **2010**, *4*, 579–585. [CrossRef]
14. Sweeney, T.; Meredith, H.; Vigors, S.; Mcdonnell, M.J.; Ryan, M.; Thornton, K.; Doherty, J.V. Extracts of laminarin and laminarin/fucoidan from the marine macroalgal species Laminaria digitata improved growth rate and intestinal structure in young chicks, but does not influence Campylobacter jejuni colonisation. *Anim. Feed Sci. Technol.* **2017**, *232*, 71–79. [CrossRef]
15. Cui, H.; Wang, Z.G.; Liu, J.; Wang, Y.X.X.; Wang, Z.X.; Fu, J.P.; Wan, Z.Y.; Li, R.D.; Li, Q.W.; Fitton, J.H.; et al. Effects of a highly purified fucoidan from Undaria pinnatifida on growth performance and intestine health status of gibel carp Carassius auratus gibelio. *Aquac. Nutr.* **2020**, *26*, 47–59. [CrossRef]
16. Sony, N.M.; Ishikawa, M.; Hossain, M.S.; Koshio, S.; Yokoyama, S. The effect of dietary fucoidan on growth, immune functions, blood characteristics and oxidative stress resistance of juvenile red sea bream, Pagrus major. *Fish Physiol. Biochem.* **2018**, *45*, 439–454. [CrossRef]
17. Liu, M.J.; Liu, W.J.; Zhang, W.J.; Yao, J.; Mo, X.C. Ultrasound-assisted extraction of boulardii yeast cell wall polysaccharides: Characterization and its biological functions on early-weaned lambs. *Food Sci. Nutr.* **2021**, *9*, 3617–3630. [CrossRef]
18. Li, Y.; Guo, Y.L.; Zhang, C.X.; Cai, X.F.; Liu, P.; Li, C.L. Effects of physical forms of starter feed on growth, nutrient digestibility, gastrointestinal enzyme activity, and morphology of pre- and post-weaning lambs. *Animal* **2020**, *15*, 100044. [CrossRef]
19. Rattigan, R.; Sweeney, T.; Vigors, S.; Thornton, K.; Rajauria, G.; O'Doherty, J.V. The Effect of Increasing Inclusion Levels of a Fucoidan-Rich Extract Derived from Ascophyllum nodosum on Growth Performance and Aspects of Intestinal Health of Pigs Post-Weaning. *Mar. Drugs* **2019**, *17*, 680. [CrossRef]
20. Draper, J.; Walsh, A.M.; McDonnell, M.; O'Doherty, J.V. Maternally offered seaweed extracts improves the performance and health status of the postweaned pig. *J. Anim. Sci.* **2016**, *94*, 391–394. [CrossRef]
21. Dang, D.X.; Liu, Y.J.; Chen, N.B.; Kim, I.H. Dietary supplementation of Aspergillus niger-expressed glucose oxidase ameliorates weaning stress and improves growth performance in weaning pigs. *J. Anim. Physiol. Anim. Nutr.* **2021**, *00*, 1–8. [CrossRef]
22. Wang, S.Q.; Ma, T.; Zhao, G.H.; Zhang, N.F.; Tu, Y.; Li, F.D.; Cui, K.; Bi, Y.L.; Ding, H.B.; Diao, Q.Y. Effect of Age and Weaning on Growth Performance, Rumen Fermentation, and Serum Parameters in Lambs Fed Starter with Limited Ewe–Lamb Interaction. *Animals* **2019**, *9*, 825. [CrossRef] [PubMed]
23. Guo, H.R.; Zhou, G.C.; Tian, G.J.; Liu, Y.Y.; Dong, N.; Li, L.F.; Zhang, S.J.; Chai, H.C.; Chen, Y.L.; Yang, Y.X. Changes in Rumen Microbiota Affect Metabolites, Immune Responses and Antioxidant Enzyme Activities of Sheep under Cold Stimulation. *Animals* **2021**, *11*, 712. [CrossRef] [PubMed]
24. Hu, H.; Bai, X.; Xu, K.X.; Zhang, C.; Chen, L. Effect of phloretin on growth performance, serum biochemical parameters and antioxidant profile in heat-stressed broilers. *Poult. Sci.* **2021**, *100*, 101217. [CrossRef] [PubMed]
25. Yang, Q.; Yang, R.; Li, M.; Zhou, Q.C.; Elmada, Z.C. Effects of dietary fucoidan on the blood constituents, anti-oxidation and innate immunity of juvenile yellow catfish (*Pelteobagrus fulvidraco*). *Fish Shellfish Immunol.* **2014**, *41*, 264–270. [CrossRef] [PubMed]
26. Zhang, Y.; Xu, M.; Hu, C.X.; Liu, A.M.; Chen, J.J.; Gu, C.F.; Zhang, X.; You, C.P.; Tong, H.B.; Wu, M.J.; et al. Sargassum fusiforme Fucoidan SP2 Extends the Lifespan of Drosophila melanogaster by Upregulating the Nrf2-Mediated Antioxidant Signaling Pathway. *Oxidative Med. Cell. Longev.* **2019**, *2019*, 8918914. [CrossRef]
27. Kim, E.T.; Lee, H.G.; Kim, D.H.; Son, J.K.; Kim, B.W.; Joo, S.S.; Park, D.S.; Park, Y.J.; Lee, S.Y.; Kim, M.H. Hydrolyzed Yeast Supplementation in Calf Starter Promotes Innate Immune Responses in Holstein Calves under Weaning Stress Condition. *Animals* **2020**, *10*, 1468. [CrossRef]

28. Shi, H.Y.; Luo, Y.R.; Li, Y.F.; Zhang, F.K.; Liu, N. Tetramethylpyrazine supplementation improves performance, digestion, blood and immune state of broilers exposure to oxidative stress. *J. Anim. Physiol. Anim. Nutr.* **2021**, *106*, 132–138. [CrossRef]
29. Chollada, B.; Sumpun, T.; Sapon, S.; Saikaew, S.; Morakot, N.; Thasinus, D.; Kazuo, K. Effects of Litter Size and Parity Number on Mammary Secretions Including, Insulin-Like Growth Factor-1, Immunoglobulin G and Vitamin A of Black Bengal, Saanen and Their Crossbred Goats in Thailand. *Vet. Sci.* **2021**, *8*, 95.
30. Xu, H.J.; Wang, L.H.; Zhang, Q.Y.; Jiang, X.; Zhang, C.R.; Zhang, Y.G. Effects of 25-hydroxyvitamin D_3 on growth performance, fecal scores, vitamin D_3 metabolites, antioxidant status, and inflammatory and stress-related parameters in weaning calves. *Anim. Feed Sci. Technol.* **2021**, *281*, 114946. [CrossRef]
31. Ying, L.Q.; Eri, R.D.; Helen, F.J.; Patel, F.J.; Nuri, G.; Britta, S. Fucoidan Extracts Ameliorate Acute Colitis. *PLoS ONE* **2015**, *10*, 128453.
32. Tomori, M.; Nagamine, T.; Miyamoto, T.; Iha, M. Evaluation of the Immunomodulatory Effects of Fucoidan Derived from Cladosiphon Okamuranus Tokida in Mice. *Mar. Drugs* **2019**, *17*, 547. [CrossRef] [PubMed]
33. Delma, C.R.; Thirugnanasambandan, S.; Srinivasan, G.P.; Raviprakash, N.; Manna, S.K.; Natarajan, M.; Aravindan, N. Fucoidan from marine brown algae attenuates pancreatic cancer progression by regulating p53—NFκB crosstalk. *Phytochemistry* **2019**, *167*, 112078. [CrossRef] [PubMed]
34. Ondrej, S.; Jakub, N.; Andrea, R.; Petr, K.; Vojtech, K.; Julius, C.; Libor, K.; Leos, P.; Lubor, L.; Eva, M. Safety of Mealworm Meal in Layer Diets and their Influence on Gut Morphology. *Animals* **2021**, *11*, 1439.
35. Deborah, A.; Fisayo, A.; Dynamics, G.M. Growth Performance, and Gut Morphology in Broiler Chickens Fed Diets Varying in Energy Density with or without Bacitracin Methylene Disalicylate (BMD). *Microorganisms* **2021**, *9*, 787.
36. Leonard, S.G.; Sweeney, T.; Bahar, B.; Lynch, B.P.; O'Doherty, J.V. Effects of dietary seaweed extract supplementation in sows and post-weaned pigs on performance, intestinal morphology, intestinal microflora and immune status. *Br. J. Nutr.* **2011**, *106*, 688–699. [CrossRef]

animals

MDPI

Article

Influence of Red Corn Rich in Anthocyanins on Productive Traits, Blood Metabolic Profile, and Antioxidative Status of Fattening Lambs

Zvonko Antunović [1], Josip Novoselec [1], Željka Klir Šalavardić [1,*], Zvonimir Steiner [1], Marcela Šperanda [1], Lidija Jakobek Barron [2], Mario Ronta [1] and Valentina Pavić [3]

1 Faculty of Agrobiotechnical Sciences Osijek, J. J. Strossmayer University of Osijek, 31000 Osijek, Croatia; zantunovic@fazos.hr (Z.A.); jnovoselec@fazos.hr (J.N.); zsteiner@fazos.hr (Z.S.); marcela.speranda@fazos.hr (M.Š.); mronta@fazos.hr (M.R.)
2 Faculty of Food and Technology Osijek, J. J. Strossmayer University of Osijek, 31000 Osijek, Croatia; lidija.jakobek@ptfos.hr
3 Department of Biology, J. J. Strossmayer University of Osijek, 31000 Osijek, Croatia; vpavic@biologija.unios.hr
* Correspondence: zklir@fazos.hr

Simple Summary: In order to prevent lamb distress during weaning and avoid the occurrence of oxidative stress leading to diminished production performance, health status, or product quality, feeds rich in polyphenols (anthocyanins) are increasingly used in ruminant feeding. In the present study, lambs were allocated into three groups, with 10 lambs per group. The feed mixture for the control group (C) contained yellow corn. Lambs in experimental group I were fed feed mixtures containing yellow corn replaced by red corn at 50% (RC50). In experimental group II, red corn fully replaced (100%) yellow corn (RC100) in the lambs' feed. The results of the present study indicate a positive effect of red corn rich in anthocyanins on the metabolic profile without any changes in the productive traits of lambs.

Abstract: In this study, we aimed to evaluate the effects of different proportions of red corn rich in anthocyanins on the diet of fattening lambs considering their productive traits, blood metabolic profile, and antioxidative status. The research was carried out with 30 Merinolandschaf lambs, 90 days old and weaned. The feed mixture for lambs (n = 10) of the control group contained yellow corn, while in the feed mixture of experimental group I (n = 10), yellow corn was replaced with red corn at 50% (RC50), and in experimental group II (n = 10), yellow corn was 100% replaced with red corn (RC100). An automatic three-part differential haematology analyser was used to determine haematological parameters in whole blood, and biochemical parameters were determined in blood serum using a biochemical analyser. A diet containing red corn did not affect productive traits or the majority of the examined parameters. However, higher blood haemoglobin content, increased aspartate aminotransferase and creatine kinase activity, and decreased glucose and non-esterified fatty acids concentrations were found in the serum of RC100 lambs. These results indicate a positive effect of red corn rich in anthocyanins on the metabolic profile without any changes in the productive traits of lambs.

Keywords: red corn; anthocyanins; lambs; productive traits; metabolic profile

Citation: Antunović, Z.; Novoselec, J.; Klir Šalavardić, Ž.; Steiner, Z.; Šperanda, M.; Jakobek Barron, L.; Ronta, M.; Pavić, V. Influence of Red Corn Rich in Anthocyanins on Productive Traits, Blood Metabolic Profile, and Antioxidative Status of Fattening Lambs. *Animals* **2022**, *12*, 612. https://doi.org/10.3390/ani12050612

Academic Editors: Panagiotis Simitzis and Athanasios I. Gelasakis

Received: 17 January 2022
Accepted: 25 February 2022
Published: 28 February 2022

1. Introduction

The weaning period is a critical stage in the rearing of young ruminants because the organism is usually under stress, which limits growth, decreases feed intake, and impairs health status [1,2]. This is due to the still poorly developed digestive system of young animals, which also contributes to possible digestive disturbances [3]. Stress in animals may occur due to inadequate nutrition and adverse environmental effects, parameters that negatively affect the intestinal microflora, leading to the development of pathogenic microorganisms which in turn cause diarrhoea, reduce nutrient absorption [4], and induce oxidative stress. Particular importance is given to improving the adaptation of young animals to concentrate feed consumption during the suckling period as soon as possible. Therefore, various feeds enriched with polyphenols, such as anthocyanins, are increasingly used as supplements incorporated into solid feed after weaning, which could lead to reduced oxidative stress and improved immune response, health status, and meat quality in lambs [5,6]. Currently, the meat industry is trying to change the consumer's perception by using natural compounds and plant-derived antioxidants instead of synthetic ones [7]. Improving diet composition is a key factor in improving the health status and welfare of animals [8] and enhancing livestock productivity [9,10]. Ruminants depend on microbial fermentation within the rumen to acquire energy from plant compounds. Rumen function has been manipulated by supplementing forages with readily fermentable carbohydrates and additives to improve animals' productivity [11,12]. Logo et al. [13] pointed out that the regulatory mechanisms of anthocyanin metabolic pathways have not yet been fully clarified. There are numerous effects of anthocyanins [14], such as antioxidative, antimicrobial, anticancer, and anti-inflammatory. Rice-Evans et al. [15] reported that anthocyanins exhibit stronger antioxidant capacity than many other antioxidants. However, Jöbstl et al. [16] observed poor palatability of anthocyanins owing to bitter taste, which may cause low digestibility and lead to a negative effect on rumen fermentation and, consequently, reduced growth of animals. Abdel-Aal et al. [17] reported that most of the anthocyanins in the coloured varieties of corn are glycosylated while some of them are acylated, such as purple and scarlet corn. It is already known that anthocyanins are primarily present in red and black coloured corn varieties, and in blue corn, they are concentrated in the aleurone [18]. In general, coloured corn varieties contain between 27 and 1439 mg anthocyanins/kg of dry kernel [19,20]. In the literature [19,20], coloured corn varieties are classified according to the average total anthocyanin contents (in mg/kg dry kernel weight), which are from 99 to 379 in blue, 26.5 to 1439 in purple, 76.2 to 120 in black, and 2.5 to 696 in pink or red corn. Žilić et al. [21] reported that red corn contains 15.43 mg of anthocyanins expressed as cyanidin-3-glucoside equivalents (CGE)/kg DM, while red-yellow and dark yellow corn contain 2.50 mg and 696.07 mg of CGE/kg DM anthocyanins, respectively. There is no information available on the use of red corn in small ruminant feeding, but there are several studies on the use of purple corn [22–24]. The conclusions from these studies are mainly related to the positive antioxidant effects of using purple corn in diets for small ruminants.

According to these findings, we postulate that replacing yellow corn with red corn in rations for lambs will not negatively affect lambs' performance, but could positively affect the metabolic profile due to increased concentrations of polyphenols or anthocyanins. Therefore, the aim of this paper is to find out how different proportions of red corn rich in anthocyanins affect the productive traits, blood metabolic profile, and antioxidative status of fattening lambs.

2. Materials and Methods

2.1. Experimental Design and Bioethics Standard

The research was carried out with 30 Merinolandschaf lambs during the fattening period at a family farm in Osijek-Baranya County, Croatia (45°20′05″ N; 18°18′59″ E). The post-weaning lambs were 90 days old, on average. The selected lambs were healthy and in satisfactory physical condition. Lambs were selected from ewes with single lambs. All lambs were dewormed at 2 months of age. These lambs were selected from 200 animals according to body weight, and were allotted to the control group, the experimental group with 50% red corn, and the experimental group with 100% red corn (25.01 ± 2.63, 25.04 ± 2.45, 25.07 ± 2.25 kg, respectively) before the experiment started. According to diet, the lambs were evenly divided by sex (50% male and 50% female) into three groups of 10 lambs. Each group was housed together in one pen (5 m × 4 m). An acclimatisation period was also carried out to adapt the lambs to the new feed, which lasted 7 days. The main experimental period lasted 27 days. Body condition score (BCS) was recorded using the 5-point scale according to Russel [25] (1 = thin to 5 = obese), and evaluated by a trained technician. Weighing and evaluation of BCS were carried out on the 1st and 27th days of the experiment. After the lambs' slaughtering and bleeding, the skin was peeled off the lambs' carcasses, and the abdominal (spleen, intestine, forestomach, stomach, and liver) and thoracic (trachea with the lungs and heart) cavity organs were removed. Afterwards, the weighing of carcasses was carried out and samples of muscle (m. semimembranosus) tissue were collected. The dressing percentage was calculated as follows: 100 × (carcass weight/live body weight).

The trial followed the recommendations of the Animal Protection Act (NN 133/06, NN 37/13 and NN 1. kg125/13), the Legal Act on the Protection of Animals Used for Scientific Purposes (NN 55/13), the European Union Directive 2010/63/EU, and the rest of the valid legal acts related to the welfare of farm animals. Therefore, the study was approved by the Bioethics Committee for Research on Animals of the Faculty of Agrobiotechnical Sciences Osijek.

2.2. Feed and Analysis of Feedstuffs

The lambs were offered the feed mixtures following their requirements (expected weight of ~32 kg) according to the National Research Council [26]. The feed mixture of lambs from the control group (C) contained only yellow corn. Yellow corn was replaced by red corn at a level of 50% in experimental group I (RC50) and at a level of 100% (RC100) in experimental group II. The red corn is an old native Croatian species. A mixture of red clover and grass hay (*Trifolium pratense* and *Lolium multiflorum*) and water were offered to lambs ad libitum. Feed mixtures were offered to the lambs twice per day. The lambs were weighed at the beginning and at the end of the study. Feed was offered at the same time each day.

All feed samples (feed mixture, hay, yellow and red corn) were dried and then ground into a powder using a heavy metal-free ultra-centrifugal mill (ZM 200, Retsch GmbH, Haan, Germany) or knife mill (GM 200, Retsch GmbH, Haan, Germany). The feed composition was determined using the standard methods of the Association of Official Analytical Chemists [27]. The ingredients and chemical compositions of the diets are presented in Table 1. The crude protein content in the feed was estimated by the Kjeldahl method using a Kjeldahl steam distillation system (Behr, Stuttgart, Germany). The ether extract was estimated by the universal extraction system B-811 (Buchi, Flawil, Switzerland). The crude fibre content was determined by the Weende method and ME was determined according to INRAE-CIRAD-AFZ [28].

The extraction of polyphenols was carried out from feed samples. First, samples were weighed and 0.2 g was set in a plastic tube, then 1.5 mL of 80% (v/v) methanol was added in water. Samples were vortexed and extracted for 15 min with an ultrasonic water bath (RK 100, Berlin, Germany), and centrifuged for 10 min at 6739× g afterwards (Eppendorf, Hamburg, Germany). The extract was transferred into a separate plastic tube. The residue was extracted again following the same procedure using methanol (0.5 mL of 80%). These two extracts were combined to obtain a final volume of the extract of around 2 mL. The same procedure was then repeated to obtain a second and third parallel feed extract.

The total concentration of feed polyphenols was determined following the Folin–Ciocalteu micro-method (Waterhouse). A diluted extract (20 µL) aliquot was mixed with distilled water (1580 µL) and Folin–Ciocalteu reagent (100 µL). An amount of 300 µL of sodium carbonate solution (200 g/L) was added to the mixture and shaken. After the incubation at 40 °C for 30 min in the water bath, the absorbance was read against the blank at 765 nm. Total polyphenols were expressed as mg of gallic acid equivalents (GAE)/kg of the sample weight. Data are presented as mean ± standard deviation of three parallels each measured two times. Determination of total anthocyanins and polyphenols was performed according to the method described by Jakobek et al. [29], using the Shimadzu UV-1280 spectrophotometer (Shimadzu Europe GmbH, Duisburg, Germany).

Table 1. The ingredients and chemical composition of the feed mixture, hay, yellow corn, and red corn used in the diets for fattening Merinolandschaf lambs.

Ingredient (g/kg Feed Mixture)	Feed Mixture	Hay [1]	Yellow Corn	Red Corn
Corn	600			
Barley	120			
Wheat flour	23			
Soybean meal (46% CP)	100			
Extruded soybean	120			
Salt	4			
Calcium carbonate	3			
Mineral–vitamin premix [2]	30			
Chemical content (g/kg DM)				
DM	912	958	905	907
Crude protein	157	96	100	105
Crude fibre (g/kg DM)	36	343	25	23
Crude ash (g/kg DM)	30	70	13	14
EE (g/kg DM)	51	9	38	37
ME (MJ/kg DM)	12.30	7	12.70	12.70
Polyphenols (total), mg/kg *	144.77	-	179.87	298.69
Anthocyanins (total), mg/kg **	0	-	125.37	253.04

[1] Red clover and grass hay (*Trifolium pratense* and *Lolium multiflorum*). [2] Mineral–vitamin premix for lambs: 8% Ca, 5% P, 9.5% Na, 2.00% Mg, 400,000 IU vitamin A, 40,000 IU vitamin D, 500 mg vitamin E, 4000 mg Zn, 2000 mg Mn, 60 mg I, 10 mg Co, 50 mg Se. CP: crude protein; DM: dry matter; EE: ether extract; ME: metabolizable energy. * Concentrations of polyphenols in RC50 and RC 100 were 513.06 and 1276.70 mg/kg, respectively. ** Concentrations of anthocyanins in RC50 and RC 100 were 217.32 and 485.40 mg/kg, respectively.

2.3. Blood Sampling and Analysis

Blood samples were collected from each lamb, at the beginning (1st day) and at the end (27th day) of the study, from the jugular vein (10 mL) into two sterile vacuum tubes (Venoject®, Sterile Terumo Europe, Leuven, Belgium) at the same time in the morning. The tubes, used for haematology analyses, contained ethylenediamine tetra-acetic acid (EDTA) as an anticoagulant. After collection, the samples were transported to the Department of Animal Production and Biotechnology (Faculty of Agrobiotechnical Sciences). The EDTA tubes were inverted several times to ensure adequate blood mixing with the anticoagulant. The automatic three-part differential haematology analyser Sysmex PocH-100iV (Sysmex Europe GmbH, Hamburg, Germany) was used to determine haematological parameters such as the number of leukocytes (WBC) and erythrocytes (RBC), the contents of haemoglobin (HGB) and haematocrit (HCT), the mean corpuscular volume (MCV), the average haemoglobin content in erythrocytes (MCH), and the mean haemoglobin concentration in erythrocytes (MCHC) in the whole blood of lambs. Afterwards, blood samples collected in sterile vacuum tubes Venoject® (Sterile Terumo Europe, Leuven, Belgium) were centrifuged at 1610× g for 10 min and the obtained serum samples were placed into the analyser Beckman Coulter AU 400 with Total Protein Reagent (Beckman Coulter Inc., Brea, CA, USA). The serum biochemical parameter concentrations were determined, such as calcium, inorganic phosphorus, magnesium, iron, urea, glucose (GUK), total proteins (PROT), albumin (ALB), cholesterol (CHOL), LDL-cholesterol (LDL), HDL-cholesterol (HDL), triglycerides (TGC), β-hydroxybutyrate (BHB), and non-esterified fatty acids (NEFA). The activities of enzymes, such as alanine aminotransferase (ALT), aspartate aminotransferase (AST), alkaline phosphatase (ALP), creatine kinase (CK), γ-glutamyl transferase (GGT), and glutathione reductase (GR), were also determined in serum. All of the above-mentioned biochemical variables were determined using Beckman Coulter reagents (Beckman Coulter Inc., Brea, CA, USA), apart from BHB concentration, which was determined by using RANBUT (Randox Laboratories Ltd., Crumlin, UK), and NEFA, determined by a NEFA kit (Randox Laboratories Ltd., Crumlin, UK). Globulin content (GLOB) was calculated as the difference between total protein and albumin. The activities of glutathione peroxidase (GPx) and superoxide dismutase (SOD) in the serum were determined using a Ransel® kit (Randox Laboratories Ltd., Crumlin, UK) and Ransod® kit (Randox Laboratories Ltd., Crumlin, UK), respectively, which were analysed by an automatic Beckman Coulter AU 400 analyser (Beckman Coulter Inc., Brea, CA, USA).

2.4. Measurements of Lipid Peroxidation and Antioxidative Activity of Meat and Serum

Immediately after slaughter, fresh lamb meat samples were collected from the right side of the m. semimembranosus, and visible fat was removed. The lambs' muscle homogenates were prepared (10% *w*/*v*) in 0.05 M phosphate buffer (pH 7) using an Ultra Turrax (IKA T18 Basic, Labortechnik, Staufen, Germany) homogeniser and centrifuged at 12,000× *g* for 60 min at 4 °C. The blood was allowed to clot, and then the serum was separated immediately by centrifugation. The meat extraction and the serum dilution preparation were carried out in triplicate. The supernatant obtained was used for the measurement of DPPH radical scavenging activity and TBARS.

2.4.1. 2,2-Diphenyl-1-picrylhydrazyl (DPPH) Radical Scavenging Activity

The total antioxidant activities of meat and serum extracts were determined using the DPPH radical scavenging assay described by Qwele et al. [30], with some modifications. The serum samples were diluted (25 µL/mL) and mixed with 0.2 mM DPPH radical solution. The muscle extracts were diluted (0.01 g mL^{-1}) and mixed with 0.2 mM DPPH radical solution. Ascorbic acid (AA) was used as a reference compound. All measurements were performed in triplicate. The absorbance was measured at 517 nm using a spectrophotometer (Lambda 25, PerkinElmer, MA, USA), and DPPH scavenging activity was determined using Equation (1):

$$\text{DPPH activity} = (A_b + A_s) - A_m)/A_b \times 100 \quad (1)$$

where A_b is the absorbance of 0.1 mM DPPH radical solution at λ = 517 nm, A_s is the absorbance of 0.1 mM extraction solution at λ = 517 nm, and A_m is the absorbance of 0.1 mM solution mixture of tested serum or extracts with DPPH radical at 517 nm.

2.4.2. Thiobarbituric Acid Reactive Substances (TBARS)

Lipid peroxidation in the meat and serum samples was estimated using the TBARS method according to Liu et al. [31], with slight modifications. A spectrophotometer (Lambda 25, PerkinElmer, MA, USA) was used to measure absorbances (532 nm and 600 nm). The molar extinction coefficient of malondialdehyde (MDA; 156,000 M^{-1} cm^{-1}) was used to calculate the MDA concentration. The results were reported as mg of MDA equivalents per kg of meat sample and as nmoles of MDA equivalents per mL of blood serum.

2.5. Statistical Analyses

The normality of data distribution was checked by the Shapiro–Wilk test (PROC UNIVARIATE of SAS) [32]. Results are expressed as mean values and the standard error of mean estimated by the MEANS procedure of SAS [32], while the effects of treatment (C-control group; RC50-50% of red corn; RC100-100% of red corn) were analysed by the GLM procedure. Tukey's test and differences between groups were declared significant at $p < 0.05$. Non-parametric data (body weight, WBC, MCV, NEFA, BHB, ALP, SOD, GPx, DPPH in blood on 27th day of study) were analysed with the Kruskal–Wallis H test.

3. Results

Table 2 shows the production traits of lambs (body weight, average daily gain, carcass dressing, and body condition scores) fed with different proportions of red corn.

Table 2. Production traits and body condition scores of weaned Merinolandschaf lambs fed various amounts of red corn.

Indicator	Day	Group (Mean)			SEM	*p*-Value
		C	RC50	RC100		
Body weight (kg)	1st 27th	24.96 32.35	25.17 32.18	25.35 33.14	0.43 0.60	0.642
Body condition score	1st 27th	3.64 3.62	3.49 3.71	3.38 3.84	0.06 0.07	0.172
Daily gain (g)	1st–27th	273.78	259.63	288.52	19.01	0.835
Carcass weight (kg)	27th	16.68	16.98	17.44	0.32	0.623
Carcass dressing (%)	27th	51.61	52.80	52.73	0.47	0.530

Mean: mean value; SEM: standard error of mean; C: control (yellow corn); RC50: red corn 50%; RC100: red corn 100%.

No significant differences were found in the productive traits of the lambs (body weight and daily gain) or in the BCS of the lambs and the carcass dressing among experimental diets (Table 2). In addition, Table 3 reports a significant increase only in blood HGB content in the RC50 and RC100 lambs compared to group C at the end of the study, while other indicators did not vary significantly among the feeding treatments.

Table 3. The haematological parameters of weaned Merinolandschaf lambs fed various amounts of red corn.

Indicator	Day	Group (Mean)			SEM	*p*-Value
		C	RC50	RC100		
White blood cells ($\times 10^9$/L blood)	1st 27th	13.09 11.04	12.68 11.53	14.78 13.26	0.81 0.68	0.653
Red blood cells ($\times 10^9$/L blood)	1st 27th	9.65 9.63	10.11 10.25	10.09 9.77	0.18 0.16	0.250
Haemoglobin (g/L blood)	1st 27th	123.90 113.00 ^{a}	124.40 124.80 ^{b}	123.80 120.40 ^{b}	2.19 1.57	0.049
Hematocrit (L/L blood)	1st 27th	0.38 0.38	0.39 0.41	0.40 0.38	0.01 0.01	0.255
Mean corpuscular volume (fL)	1st 27th	39.31 39.87	39.38 39.88	39.07 38.74	0.50 0.49	0.512
Mean corpuscular haemoglobin (pg)	1st 27th	12.94 11.76	12.33 12.42	12.30 11.49	0.23 0.41	0.651
Mean corpuscular haemoglobin concentration (g/L blood)	1st 27th	331.20 297.60	315.50 308.50	317.10 321.60	8.17 7.63	0.452
Platelet blood count ($\times 10^9$/L blood)	1st 27th	741.00 536.10	561.70 508.00	744.10 599.10	39.41 48.43	0.747

Mean: mean value; SEM: standard error of mean; C: control (yellow corn); RC50: red corn 50%; RC100: red corn 100%. ^{a,b} Values in rows with different letters differ significantly ($p < 0.05$).

There were no significant differences in most biochemical parameters in the blood serum of lambs fed RC50 and RC100, except for glucose and NEFA concentrations (Table 4). The blood glucose and NEFA concentrations of RC100 lambs were significantly reduced at the end of the study compared to group C.

Table 4. The biochemical parameters in the blood serum of weaned Merinolandschaf lambs fed various amounts of red corn.

Indicator	Day	Group (Mean)			SEM	*p*-Value
		C	RC50	RC100		
Mg (mmol/L)	1st 27th	1.41 1.36	1.38 1.34	1.43 1.33	0.02 0.02	0.856
Ca (mmol/L)	1st 27th	2.62 2.61	2.50 2.74	2.55 2.65	0.05 0.03	0.103
Fe (μmol/L)	1st 27th	33.39 26.82	34.94 33.42	37.13 32.62	1.52 1.76	0.257
P-inorganic (mmol/L)	1st 27th	2.79 2.72	2.91 2.87	2.98 2.99	0.09 0.09	0.461
Glucose (mmol/L)	1st 27th	5.51 5.98 [a]	5.09 5.74 [a]	5.40 5.27 [b]	0.12 0.09	0.003
Urea (g/L)	1st 27th	4.92 5.41	3.52 5.60	4.51 5.41	0.26 0.17	0.874
Total protein (g/L)	1st 27th	60.65 63.35	58.38 64.030	58.90 63.06	0.51 0.73	0.865
Albumin (g/L)	1st 27th	29.90 29.51	29.92 31.00	30.50 30.75	0.29 0.31	0.109
Globulin (g/L)	1st 27th	30.75 33.84	28.46 33.03	28.40 32.31	0.46 0.66	0.654
Cholesterol (mmol/L)	1st 27th	1.97 1.33	1.96 1.28	1.93 1.21	0.10 0.05	0.196
Triglycerides (mmol/L)	1st 27th	0.34 0.23	0.36 0.22	0.36 0.18	0.03 0.02	0.532
HDL (mmol/L)	1st 27th	1.05 0.73	1.02 0.73	1.05 0.74	0.04 0.03	0.768
LDL (mmol/L)	1st 27th	0.80 0.47	0.74 0.39	0.71 0.44	0.06 0.03	0.188
NEFA (mmol/L)	1st 27th	0.33 0.31 [a]	0.23 0.17 [ab]	0.40 0.08 [b]	0.05 0.04	0.013
Beta hydroxybutyrate (mmol/L)	1st 27th	0.35 0.43	0.37 0.51	0.40 0.42	0.03 0.03	0.193

Mean: mean value; SEM: standard error of mean; C: control (yellow corn); RC50: red corn 50%; RC100: red corn 100%; Mg: magnesium; Ca: calcium; Fe: iron; P-inorganic: phosphorus; NEFA: non-esterified fatty acids. [a,b] Values in rows with different letters differ significantly ($p < 0.05$).

The analysis of enzyme activity in the blood serum of lambs fed with different RC proportions revealed a significant effect, especially a significant increase in CK and AST activity in RC100 lambs compared to the C group (Table 5). Other determined enzyme activities did not differ significantly among the different dietary treatments.

Table 5. The activity of enzymes in the blood serum of weaned Merinolandschaf lambs fed various amounts of red corn.

Enzymes	Day	Group (Mean)			SEM	*p*-Value
		C	RC50	RC100		
Aspartate transaminase (AST, U/L)	1st 27th	133.43 87.96 [a]	97.59 94.30 [ab]	106.76 104.54 [b]	11.64 2.54	0.021
Alanine aminotransferase (ALT, U/L)	1st 27th	8.76 8.82	11.22 8.35	9.42 8.86	1.01 0.43	0.875
Alkaline phosphatase (ALP, U/L)	1st 27th	350.68 358.30	366.26 307.48	490.17 456.97	31.54 28.28	0.119
Gamma-glutamyl transferase (GGT, U/L)	1st 27th	96.02 79.96	77.25 81.23	76.47 74.02	5.97 2.30	0.408
Creatine kinase (CK, U/L)	1st 27th	160.40 119.80 [a]	274.20 127.80 [ab]	184.10 162.20 [b]	21.15 7.41	0.039
Superoxide dismutase (SOD, U/mL)	1st 27th	0.38 0.55	0.50 0.61	0.42 0.66	0.08 0.05	0.264
Glutathione reductase (GR, U/L)	1st 27th	72.35 80.44	75.91 83.65	72.26 93.18	3.16 4.07	0.429
Glutathione peroxidase (GPx, U/L)	1st 27th	522.60 582.30	441.22 626.30	491.80 625.10	33.13 57.35	0.882

Mean: mean value; SEM: standard error of mean; C: control (yellow corn); RC50: red corn 50%; RC100: red corn 100%. [a,b] Values in rows with different letters differ significantly ($p < 0.05$).

The analysis of antioxidant status reveals no significant differences in TBARS or DPPH in blood serum or in the meat samples of lambs regardless of dietary treatment (Table 6).

Table 6. Antioxidative status of lambs' blood serum and meat.

Parameter	Day	Group (Mean)			SEM	*p*-Value
		C	RC50	RC100		
TBARS (nmol MDA/mL blood serum)	1st 27th	0.72 0.84	0.71 0.83	0.64 0.73	0.03 0.04	0.370
DPPH blood serum (%) *	1st 27th	73.63 73.07	77.48 74.33	76.21 74.20	2.30 3.17	0.983
TBARS (mg MDAeq/kg muscle)		4.21	4.10	3.70	0.17	0.445
DPPH muscle (%) **		54.10	56.84	57.08	1.01	0.433

Mean: mean value; SEM: standard error of mean; C: control (yellow corn); RC50: red corn 50%; RC100: red corn 100%; TBARS: thiobarbituric acid reactive substances. * % DPPH radical scavenging activity at final serum concentration 25 µL/mL; ** % DPPH radical scavenging activity at final muscle concentration 0.01 g/mL.

4. Discussion

The weaning period represents a stressful condition for animals. In particular, in ruminants, weaning is not just a single event but a period in which milk is progressively replaced by forage and concentrate or by grain-based diets. Lambs are not ruminants at birth, and during the suckling period, the animal's forestomach is poorly developed. As dry feeds begin to be consumed, the rumen and the reticulum begin to develop and rapidly increase in size. It has been demonstrated that small ruminants are more prone to oxidative stress due to intense metabolic requirements. The appearance of oxidative stress in ruminants endangers their productivity and their health. Therefore, research on the addition of polyphenols, including anthocyanins, in the diet of small ruminants is common, given the various biological activities of polyphenols (particularly anthocyanins) in different corn varieties [33]. In the present study, there were no significant changes in the productive traits of lambs, such as body weight and daily weight gain, or in the BCS and carcass dressing among the different feeding treatments. However, it is noticeable that the expected body weight and daily gain of the lambs were achieved in all groups (Table 2). In previous works carried out with the addition of polyphenols in lambs' diet, the body weight was not affected by grapeseed oil or grapeseed extract addition [34,35]. Kanfantaris et al. [36] concluded that the supplementation of polyphenols in lamb diets, such as grape pomace, is more effective in improving production traits during the suckling period than in weaned lambs.

In the present study, haematological and biochemical parameters as well as parameters of enzyme activity in the lambs' blood ranged within the reference values established for lambs reared in similar conditions [37–40]. An increase in the haemoglobin content in the blood of the lambs of the RC50 and RC100 groups compared to the C lambs was found. This may be related to anthocyanins that stabilise erythrocyte membranes and inhibit haemoglobin polymerisation [41]. It is well known that polyphenols protect against reactive oxygen species-induced haemolysis via increased red blood cell integrity associated with the inhibition of lipid peroxidation [42]. The concentration of glucose and NEFA in the blood serum of the RC100 lambs was significantly reduced at the end of the study compared to group C. Glucose concentrations in the blood serum of the lambs in the RC50 group were reduced at the end of the study, although not significantly when compared to the C group, but it was significant compared to the RC100 lambs. There was a slight numerical decrease in the blood serum CHOL and TG concentrations of the RC50 and RC100 lambs compared to the C lambs. These changes indicate the need for a longer experimental duration when changes in glucose and NEFA concentrations would be evident with certainty. Sharma et al. [43] carried out a 12-week trial with mice fed diets with a high content of fat supplemented with isocaloric white, purple, or black whole wheat. These authors discovered that body weight gain was significantly reduced in mice fed black wheat. In contrast, both black and purple varieties of wheat reduced serum cholesterol concentrations, triglycerides, and free fatty acids while restoring normal serum glucose concentrations and insulin resistance. Chen et al. [44] carried out a study with mice fed black soybean seed coat extract (BSSCE), a rich source of anthocyanins. A significant decrease in serum glucose, cholesterol, triglyceride, and NEFA and MDA concentrations was determined, as well as an increase in serum HDL-cholesterol concentration and antioxidant enzyme (SOD, GPx, and catalase) activities. The authors emphasised that cyanidin-3-O-glucoside contributed to the BSSCE-induced hypoglycaemia and hypolipidemia in type 2 diabetes mellitus. Tian et al. [45] found a significant reduction in total blood cholesterol during the 74 days of their study when feeding goat kids a diet supplemented with 0.5 g/d or 1 g/d anthocyanin-rich purple corn pigment. The activity of AST in serum, GGT, ALP, and cholesterol concentrations are used for hepatic damage diagnosis in humans and animals [46], while ALP activity and cholesterol concentrations are used to detect bile obstruction or mild and progressive liver damage. A liver enzyme such as ALT, a specific hepatocellular enzyme released after hepatocellular damage, more than GGT, is used to assess liver damage. Despite significant differences in AST and CK at the end of the present study, it can be concluded that since their activities were within the reference values [47], no damage to the liver and muscle occurred. SOD, GPx, and catalase are all antioxidant enzymes that help maintain a healthy cellular antioxidant status [48]. The absence of significant changes in antioxidant enzyme activity (SOD, GPx, GR) might be explained by the lower absorption of anthocyanin by small ruminants compared to non-ruminant animals [49]. Hosoda et al. [22] reported that one of the reasons could be a sufficient pool of non-enzymatic antioxidants since severe oxidative stress is not determined in sheep. In a study conducted with lactating sheep fed purple corn, these authors found increased plasma SOD activity. However, due to the lack of research with red corn, the present study results were compared with studies in which purple corn was used, which contains a significantly higher content of anthocyanins than red corn [21]. The anthocyanin concentration possibly was not high enough to improve the antioxidative status based on SOD, GPx, or GR. A relatively short duration of the present study also contributed to an absence of significant differences.

5. Conclusions

The administration of feed mixtures containing red corn rich in anthocyanins affected neither the productive traits nor most of the haematological and biochemical indicators of blood in the lambs studied here. However, significantly higher blood HGB content and increases in serum AST and CK activity were found in the RC100 lambs, as well as decreases in the serum GUK and NEFA concentrations compared to the C group. For further research, it is necessary to start earlier in the suckling period and extend the duration of fattening, with the inclusion in the experimental model the analysis of other muscles and qualitative properties of meat in order to more comprehensively observe the response to red corn rich in anthocyanins used in lamb diets.

Author Contributions: Conceptualisation, Z.A.; investigation, J.N., Ž.K.Š., M.R., Z.S.; writing—original draft preparation, Z.A.; writing—review and editing, Z.A., Ž.K.Š., V.P.; supervision, Z.A., V.P.; formal analysis, M.Š., L.J.B., V.P.; visualisation, Z.S. All authors have read and agreed to the published version of the manuscript.

Funding: This research received no external funding.

Institutional Review Board Statement: The animal study protocol was approved by the Bioethical Committee for Research on Animals of Faculty of Agrobiotechnical Science Osijek (21-06, 22 December 2021).

Data Availability Statement: The data are available upon request.

Acknowledgments: The research was carried out with the Innovative Breeding and Technological Processes in Animal Production (no. 1126) research team at the Faculty of Agrobiotechnical Sciences Osijek.

Conflicts of Interest: The authors declare no conflict of interest.

References

1. Antunović, Z.; Šperanda, M.; Senčić, Đ.; Domaćinović, M.; Novoselec, J. Djelotvornost probiotskog pripravka "Probios 2B" u hranidbi jaradi. *Krmiva* **2008**, *50*, 73–78.
2. Lucianer, E.; da Silva, A.S.; Cazarotto, C.J.; Alba, D.F.; Griss, L.G.; Zampar, A.; Vedovatto, M.; Kessler, J.D. Addition of tannin in lamb diet after weaning: Impact on performance and hematological and biochemical variables. *Acta Acientiae Vet.* **2019**, *47*, 1–11. [CrossRef]
3. Boas, A.S.V.; Arrigoni, M.B.; Silveira, A.C.; Costa, C.; Chardulo, L.A.L. Effects of age at weaning and feed management on the production of super-young lambs. *Rev. Bras. De Zootech.* **2003**, *32*, 1969–1980.
4. Hirsh, D.C. Relation of normal microflora to structures and functions of the gastrointestinal tract. In *Veterinary Gastroenterology*; Anderson, N.V., Ed.; Lea & Febiger: Philadelphia, PA, USA, 1980; p. 207.
5. Correddu, F.; Lunesu, M.F.; Buffa, G.; Atzori, A.S.; Nudda, A.; Battacone, G.; Pulina, G. Can agro-industrial by-products rich in polyphenols be advantageously used in the feeding and nutrition of dairy small ruminants? *Animals* **2020**, *10*, 131. [CrossRef]
6. Simitzis, P.E.; Charismiadou, M.A.; Goliomytis, M.; Charalambous, A.; Ntetska, I.; Giamouri, E.; Deligeorgis, S.G. Antioxidant status, meat oxidative stability and quality characteristics of lambs fed with hesperidin, naringin or alpha-tocopheryl acetate supplemented diets. *J. Sci. Food Agric.* **2019**, *99*, 343–349. [CrossRef] [PubMed]
7. Manessis, G.; Kalogianni, A.I.; Lazou, T.; Moschovas, M.; Bossis, I.; Gelasakis, A.I. Plant-derived natural antioxidants in meat and meat products. *Antioxidants* **2020**, *9*, 1215. [CrossRef]
8. Abbate, J.M.; Macrì, F.; Capparucci, F.; Iaria, C.; Briguglio, G.; Cicero, L.; Salvo, A.; Arfuso, F.; Ieni, A.; Piccione, G.; et al. Administration of protein hydrolysates from anchovy (*Engraulis Encrasicolus*) waste for twelve weeks decreases metabolic dysfunction-associated fatty liver disease severity in ApoE−/−Mice. *Animals* **2020**, *10*, 2303. [CrossRef]
9. Avondo, M.; Pagano, R.; Guastella, A.; Criscione, A.; di Gloria, M.; Valenti, B.; Piccione, G.; Pennisi, P. Diet selection and milk production and composition in Girgentana goats with different αs1-casein genotype. *J. Dairy Res.* **2009**, *76*, 202–209. [CrossRef]
10. Monteverde, V.; Congiu, F.; Vazzana, I.; Dara, S.; di Pietro, S.; Piccione, G. Serum lipid profile modification related to polyunsaturated fatty acid supplementation in thoroughbred horses. *J. Appl. Anim. Res.* **2017**, *45*, 615–618. [CrossRef]
11. Gobindram, M.N.N.E.; Bognanno, M.; Luciano, G.; Avondo, M.; Piccione, G.; Biondi, L. The effects of barley replacement by dehydrated citrus pulp on feed intake, performance, feeding behaviour and serum metabolic indicators in lambs. *Anim. Prod. Sci.* **2017**, *57*, 133–140. [CrossRef]
12. Armato, L.; Gianesella, M.; Morgante, M.; Fiore, E.; Rizzo, M.; Giudice, E.; Piccione, G. Rumen volatile fatty acids × dietary supplementation with live yeast and yeast cell wall in feedlot beef cattle. *Acta Agric. Scand. Sect. A-Anim. Sci.* **2016**, *66*, 119–124. [CrossRef]

13. Logo, C.; Cassani, E.; Petroni, K.; Calvenzani, V.; Tonelli, C.; Pilu, R. Study of maize genotypes rich in antocyanins for human and animal nutrition. In Proceedings of the Joint Meeting AGI-SIBV-SIGA, Assisi, Italy, 19–22 September 2011.
14. Changxing, L.; Chenling, M.; Alagawany, M.; Jinhua, L.; Dongfand, D.; Gaichao, W.; Wenyin, Z.; Syed, S.F.; Arain, M.A.; Saeed, M.; et al. Health benefits and potential applications of anthocyanins in poultry feed industry. *Worlds Poult. Sci. J.* **2018**, *74*, 251–263. [CrossRef]
15. Rice-Evans, C.A.; Miller, N.J.; Bolwell, P.G.; Bramley, P.M.; Pridham, J.B. The relative antioxidant activities of plant-derived polyphenolic flavonoids. *Free Radic. Res.* **1995**, *22*, 375–383. [CrossRef]
16. Jöbstl, E.; O'Connell, J.; Fairclough, J.P.; Williamson, M.P. Molecular model for astringency produced by polphenol/protein interactions. *Biomacromolecules* **2004**, *5*, 942–949. [CrossRef] [PubMed]
17. Abdel-Aal, E.S.M.; Young, J.C.; Rabalski, I. Anthocyanin composition in black, blue, pink, purple, and red cereal grains. *J. Agric. Food Chem.* **2006**, *54*, 4696–4704. [CrossRef]
18. Chatham, L.A.; Paulsmeyer, M.; Juvik, J.A. Prospects for economical natural colorants: Insights from maize. *Theor. Appl. Genet.* **2019**, *132*, 2927–2946. [CrossRef]
19. Salinas, M.Y.; Sanchez, G.S.; Hernandez, D.R.; Lobato, N.R. Characterization of anthocyanin extracts from maize kernels. *J. Chromatogr. Sci.* **2005**, *43*, 483–487. [CrossRef]
20. Lopez-Martinez, L.X.; Oliart-Ros, R.M.; Valerio-Alfaro, G.; Lee, C.H.; Parkin, K.L.; Garcia, H.S. Antioxidant activity, phenolic compounds and anthocyanins content of eighteen strains of Mexican maize. *LWT-Food Sci. Technol.* **2009**, *42*, 1187–1192. [CrossRef]
21. Žilić, S.; Serpen, A.; Akillioglu, G.; Gökmen, V.; Vančcetović, J. Phenolic compounds, carotenoids, anthocyanins, and antioxidant capacity of colored maize (*Zea mays* L.) kernels. *J. Agric. Food Chem.* **2012**, *60*, 1224–1231. [CrossRef]
22. Hosoda, K.; Miyaji, M.; Matsuyama, H.; Haga, S.; Ishizaki, H.; Nonaka, K. Effect of supplementation of purple pigment from anthocyanin-rich corn (*Zea mays* L.) on blood antioxidant activity and oxidation resistance in sheep. *Livest. Sci.* **2012**, *145*, 266–270. [CrossRef]
23. Tian, X.Z.; Paengkoum, P.; Paengkoum, S.; Chumpawadee, S.; Ban, C.; Sorasak, T. Purple corn (*Zea mays* L.) stover silage with abundant anthocyanins transferring anthocyanin composition to the milk and increasing antioxidant status of lactating dairy goats. *J. Dairy Sci.* **2019**, *102*, 413–418. [CrossRef] [PubMed]
24. Tian, X.Z.; Xin, H.; Paengkoum, P.; Paengkoum, S.; Ban, C.; Sorasak, T. Effects of anthocyanin-rich purple corn (*Zea mays* L.) stover silage on nutrient utilization, rumen fermentation, plasma antioxidant capacity, and mammary gland gene expression in dairy goats. *J. Anim. Sci.* **2019**, *97*, 1384–1397. [CrossRef] [PubMed]
25. Russel, A. Body condition scoring of sheep. In *Sheep and Goat Practice*; Boden, E., Ed.; Bailliere Tindall: Philadelphia, PA, USA, 1991; p. 3.
26. National Research Council (NRC). *Nutrient Requirements of Small Ruminants: Sheep, Goats, Cervids and New World Camelids*; The National Academy Press: Washington DC, WA, USA, 2007; p. 256.
27. Association of Official Analytical Chemists (AOAC). *Official Methods of Analysis*, 18th ed.; AOAC: Arlington, VA, USA, 2006; pp. 24–43.
28. INRAE-CIRAD-AFZ. Feed Tables Composition and Nutritive Values of Feeds for Cattle, Sheep, Goats, Pigs, Poultry, Rabbits, Horses and Salmonids. Available online: https://www.feedtables.com (accessed on 20 December 2021).
29. Jakobek, L.; Matić, P.; Ištuk, J.; Barron, A.R. Study of interactions between individual phenolics of aronia with barley β-glucan. *Pol. J. Food Nutr. Sci.* **2021**, *71*, 187–196. [CrossRef]
30. Qwele, K.; Hugo, A.; Oyedemi, S.O.; Moyo, B.; Masika, P.J.; Muchenje, V. Chemical composition, fatty acid content and antioxidant potential of meat from goats supplemented with Moringa (*Moringa oleifera*) leaves, sunflower cake and grass hay. *Meat Sci.* **2013**, *93*, 455–462. [CrossRef] [PubMed]
31. Liu, F.; Dai, R.; Zhu, J.; Li, X. Optimizing color and lipid stability of beef patties with a mixture design incorporating with tea catechins, carnosine, and α-tocopherol. *J. Food Eng.* **2010**, *98*, 170–177. [CrossRef]
32. *SAS® 9.4 2002.-2012*; SAS Institute Inc., SAS Campus Drive: Cary, NC, USA, 2012.
33. Colombo, R.; Ferron, L.; Papetti, A. Colored Corn: An up-date on metabolites extraction, health implication, and potential use. *Molecules* **2021**, *26*, 199. [CrossRef]
34. Sharifi, M.; Bashtani, M.; Naserian, A.A.; Farhangfar, H.; Rasani, M.; Emami, A. Grape seed oil supplementation in lamb diet: Effect on meat oxidation stability and muscle fatty acids. *Ital. J. Anim. Sci.* **2019**, *18*, 1302–1309. [CrossRef]
35. Giller, K.; Sinza, S.; Messadene-Chelalib, J.; Marquardt, S. Maternal and direct dietary polyphenol supplementation affect growth, carcass and meat quality of sheep and goats. *Animal* **2021**, *15*, 100333. [CrossRef]
36. Kafantaris, I.; Kotsampasi, B.; Christodoulou, V.; Makri, S.; Stagos, D.; Gerasopoulos, K.; Petrotos, K.; Goulas, P.; Kouretas, D. Effects of dietary grape pomacesupplementation on performance, carcass traits and meat quality of lambs. *In Vivo* **2018**, *32*, 807–812. [CrossRef]
37. Latimer, K.S.; Prasse, K. Leukocites. In *Duncan and Prasse's Veterinary Laboratory Medicine—Clinical Pathology*, 4th ed.; Latimer, K.S., Maheffey, E.A., Prasse, K.W., Eds.; Iowa State University Press: Iowa City, IA, USA, 2003; pp. 46–79.
38. Lepherd, M.; Canfield, P.J.; Hunt, G.B.; Bosward, K.L. Haematological, biochemical and selected acute phase protein reference intervals for weaned female Merino lambs. *Aus. Vet. J.* **2009**, *87*, 5–11. [CrossRef]

39. Antunović, Z.; Novoselec, J.; Senčić, Đ.; Šperanda, M.; Steiner, Z.; Samac, D. Production traits and blood biochemical parameters in organic lamb production. In Proceedings of the 45th Croatian and 5th International Symposium on Agriculture, Opatija, Croatia, 15–19 February 2010.
40. Antunović, Z.; Mioč, B.; Klir Šalavardić, Ž.; Širić, I.; Držaić, V.; Đidara, M.; Novoselec, J. The effect of lactation stage on the hematological and biochemical parameters of the Travnik Pramenka ewes. *Poljoprivreda* **2021**, *27*, 56–62. [CrossRef]
41. Sivamaruthi, B.S.; Kesika, P.; Chaiyasut, C. The influence of supplementation of anthocyanins on obesity-associated comorbidities: A concise review. *Foods* **2020**, *9*, 687. [CrossRef]
42. Youdim, K.A.; Shukitt-Hale, B.; MacKinnon, S.; Kalt, W.; Joseph, J.A. Polyphenolics enhance red blood cell resistance to oxidative stress: In vitro and in vivo. *Biochim. Biophys. Acta* **2000**, *1523*, 117–122. [CrossRef]
43. Sharma, S.; Khare, P.; Kumar, A.; Chunduri, V.; Kumar, A.; Kapoor, P.; Mangal, P.; Kondepudi, K.K.; Bishnoi, M.; Garg, M. Anthocyanin-biofortified colored wheat prevents high fat diet–induced alterations in mice: Nutrigenomics studies. *Mol. Nutr. Food Res.* **2020**, *64*, 1–12.
44. Chen, Z.; Wang, C.; Pan, Y.; Gao, X.; Chen, H. Hypoglycemic and hypolipidemic effects of anthocyanins extract from black soybean seed coat in high fat diet and streptozotocin-induced diabetic mice. *Food Funct.* **2018**, *24*, 426–439. [CrossRef]
45. Tian, X.; Lu, Q.; Zhao, S.; Li, J.; Luo, Q.; Wang, X.; Zhang, Y.; Zheng, N. Purple corn anthocyanin affects lipid mechanism, flavor compound profiles, and related gene expression of longissimus thoracis et lumborum muscle in goats. *Animals* **2021**, *11*, 2407. [CrossRef] [PubMed]
46. Silanikove, N.; Tiomokin, D. Toxicity induced by poultry litter consumption: Effect on parameters reflecting liver function in beef cows. *Anim. Prod.* **1992**, *54*, 203–209. [CrossRef]
47. Kaneko, J.J.; Harvey, J.W.; Bruss, M.L. *Clinical Biochemistry of Domestic Animals*, 6th ed.; Elsevier/Academic Press: Amsterdam, The Netherlands, 2008; p. 963.
48. Peng, K.; Shirley, D.C.; Xu, Z.; Huang, Q.; McAllister, T.A.; Chaves, A.V.; Acharya, S.; Liu, C.; Wang, S.; Wang, Y. Effect of purple prairie clover (*Dalea purpurea Vent.*) hay and its condensed tannins on growth performance, wool growth, nutrient digestibility, blood metabolites and ruminal fermentation in lambs fed total mixed rations. *Anim. Feed Sci. Technol.* **2016**, *222*, 100–110. [CrossRef]
49. Dijkstra, J.; Forbes, J.M.; France, J. Introduction. In *Quantitive Aspects of Ruminant Digestion and Metabolism*, 2nd ed.; Dijkstra, J., Forbes, J.M., France, J., Eds.; CABI Publishing: Wallingford, UK, 2005; pp. 1–10.

animals

MDPI

Article

Introducing Mediterranean Lupins in Lamb Diets: Effects on Carcass Composition, Meat Quality, and Intramuscular Fatty Acid Profile

Mariana Almeida [1,2,3,*], Sofia Garcia-Santos [4], Daniela Carloto [3], André Arantes [3], Jose M. Lorenzo [5,6], José António Silva [1,2,3], Virgínia Santos [1,2,3], Jorge Azevedo [1,2,3], Cristina Guedes [1,2,3], Luís Ferreira [3,4] and Severiano Silva [1,2,3]

1 Veterinary and Animal Research Centre (CECAV), University of Trás-os-Montes e Alto Douro, Quinta de Prados, 5000-801 Vila Real, Portugal; jasilva@utad.pt (J.A.S.); vsantos@utad.pt (V.S.); jazevedo@utad.pt (J.A.); cguedes@utad.pt (C.G.); ssilva@utad.pt (S.S.)

2 Associate Laboratory for Animal and Veterinary Sciences (AL4AnimalS), University of Trás-os-Montes e Alto Douro, Quinta de Prados, 5000-801 Vila Real, Portugal

3 Department of Animal Science, University of Trás-os-Montes e Alto Douro, Quinta de Prados, 5000-801 Vila Real, Portugal; daniela30121996@hotmail.com (D.C.); arantex@hotmail.com (A.A.); lmf@utad.pt (L.F.)

4 Centre for the Research and Technology Agro-Environmental and Biological Sciences (CITAB), University of Trás-os-Montes e Alto Douro, Quinta de Prados, 5000-801 Vila Real, Portugal; ssantos@utad.pt

5 Centro Tecnológico de la Carne de Galicia, Rúa Galicia N°4, Parque Tecnológico de Galicia, San Cibrán das Viñas, 32900 Ourense, Spain; jmlorenzo@ceteca.net

6 Área de Tecnología de los Alimentos, Facultad de Ciencias de Ourense, Universidad de Vigo, 32004 Ourense, Spain

* Correspondence: mdantas@utad.pt

Citation: Almeida, M.; Garcia-Santos, S.; Carloto, D.; Arantes, A.; Lorenzo, J.M.; Silva, J.A.; Santos, V.; Azevedo, J.; Guedes, C.; Ferreira, L.; et al. Introducing Mediterranean Lupins in Lamb Diets: Effects on Carcass Composition, Meat Quality, and Intramuscular Fatty Acid Profile. *Animals* **2022**, *12*, 1758. https://doi.org/10.3390/ani12141758

Academic Editors: Panagiotis Simitzis and Athanasios I. Gelasakis

Received: 20 May 2022
Accepted: 7 July 2022
Published: 8 July 2022

Simple Summary: The main aim of this preliminary study was to evaluate the effects of replacing soybean meal with lupins on carcass traits, meat characteristics, meat characteristics, and meat fatty acid profile in lambs. Two trials were conducted: In trial 1, the soybean meal was partially replaced by *Lupinus albus* or *Lupinus luteus*; in trial 2, lambs were fed four diets with graded levels of *Lupinus luteus*, ranging from 0 to 200 g/kg. The lambs were slaughtered to evaluate carcass characteristics, meat composition, and fatty acids profile. Carcass composition was not affected ($p > 0.05$) by diet in both trials. Meat quality attributes did not vary ($p < 0.05$) between trials 1 and 2. Overall, fatty acid content was not affected by diet ($p > 0.05$) in both trials. Soybean meal produced the same results as lupins in this study, indicating the latter as a potential alternative protein source, although research should focus on meat palatability.

Abstract: The objective of this preliminary study was to evaluate the effects of partial replacement of soybean meal by lupins on lambs' diets, on the carcass traits, meat characteristics, and meat fatty acid profile. Two trials were conducted: In trial 1, the soybean meal (control; C) was partially replaced by *Lupinus albus* or *Lupinus luteus* (50 g/kg; LA5 and LL5, respectively); in trial 2, lambs were fed four diets with graded levels of *Lupinus luteus* (0, 100, 150 and 200 g/kg; C, LL10, LL15, LL20, respectively). At the end of the feeding trials, animals were slaughtered to evaluate carcass characteristics and meat composition, including fatty acids. Carcass composition in tissues was not affected ($p > 0.05$) by diet in both trials. Additionally, no significant ($p < 0.05$) differences were observed in meat quality attributes between diets in trials 1 and 2. Overall, the *Longissimus* muscle's fatty acid content was not affected by diet ($p > 0.05$) in both trials. Carcass and meat quality was overall comparable between lambs fed with soybean meal and lupins, indicating the latter as a potential alternative protein source. However, the lack of significant differences could also be attributed to the small sample size.

Keywords: lupins; lambs; meat quality; pH; color; carcass characteristics; fatty acids

1. Introduction

Over the last decade, the European Union (EU) has focused on solving its dependency on imported soybean for animal feeding [1]. Different protein alternatives have been evaluated, such as faba beans (*Vicia faba*), peas (*Pisum sativum*), and lupins (*Lupinus* spp.). These species are well-adapted to the Mediterranean climate and soil characteristics [1,2]. Although these Mediterranean legumes species offer lower protein contents than soybean meal, which is the most common protein source [3], they represent a local solution and possible replacement candidate for soybean meal in livestock feeding [4,5]. The inclusion of lupins as an alternative protein source in ruminant feeding has shown positive results in ewe milk production and composition, and it improved nursing performance [6,7], resulting in similar values to those obtained with soybean meal diets. Most important are the results found on carcass traits and meat quality of lambs fed with lupins [8–10], which are very encouraging and again indicate that this might be an adequate substitute. Facciolongo et al. [8] found comparable results on carcass traits between Awassi lambs fed soybean meal and lupins, which is in accordance with more recent results found in Gentile di Puglia lambs [9]. Overall, the inclusion of legume grains in ruminant feeding, lupins in specific, usually provides very similar results in meat quality and carcass traits to the ones obtained with soybean meal, as previously stated. Most studies, however, present results on the influence of *Lupinus albus* (white lupin) incorporation on ruminant diets, and few studies on *Lupinus luteus* (yellow lupins) incorporation in finishing lambs' diets have been published, although this inclusion has provided comparable results to those from a soybean-meal-based diet [11]. Although there has been a decline in lamb meat production and consumption in Portugal during the last decade [12], sheep are extremely well-adapted to the Mediterranean diverse production systems and could benefit from replacing soybean meal with lupins [13], especially yellow lupins which used to be cultivated in the area. This study provides results on the inclusion of yellow lupins in lambs' diets, which is not common and could provide a relevant step toward broader studies with legume grains. After analyzing the inclusion of *Lupinus albus* and *Lupinus luteus* in *Churra da Terra Quente* lambs' diets and its effects on performance [14], this study aimed to evaluate the effects of this inclusion on carcass composition and meat quality.

2. Materials and Methods

Two different trials were conducted over two consecutive years to study the effect of introducing Mediterranean lupins in diets on lamb's carcass composition and meat quality. The first trial evaluated the introduction of low percentages of *Lupinus albus* and *Lupinus luteus*, while the second one focused on increasing levels of *Lupinus luteus*. Both trials were held at the University of Trás-os-Montes and Alto Douro (UTAD) at Vila Real (Portugal). Daily handling was performed by trained personnel while respecting the Portuguese law on animal welfare in experimental research [15]. The protocol was approved by the ORBEA (Animal Welfare Body) of the University (669-e-DZ-2018).

2.1. Animals and Housing

A total of 28 Churra da Terra Quente weaned male lambs were used in the trials. In the first trial, 12 lambs with ages between 92 and 110 days and initial body weight (BW) of 18 $\pm$ 2.8 kg were distributed between 3 groups of 4 animals each. In trial 2, 16 more animals were added, so a total of 16 lambs (16 $\pm$ 2.6 kg BW and 92–110 days of age) were split into 4 groups of 4 animals each. Both trials started 3 weeks after weaning a 21-day adaptation period and lasted 12–16 weeks.

2.2. Diets

During the trials, each group received a different diet (Figure 1). Three diets were provided in trial 1: a control diet (C; 150 g/kg soybean meal) without lupin incorporation, a diet with 50 g/kg *Lupinus luteus* cv. Mister (LL5), and a diet with 50 g/kg *Lupinus albus* cv. Nacional (LA5). In trial 2, the lambs were fed four different diets: one group was provided with the control diet (C; 170 g/kg soybean meal), and the others consumed diets with different incorporations of *Lupinus luteus* cv. Mister (100 g/kg, 150 g/kg, 200 g/kg; LL10, LL15, LL20).

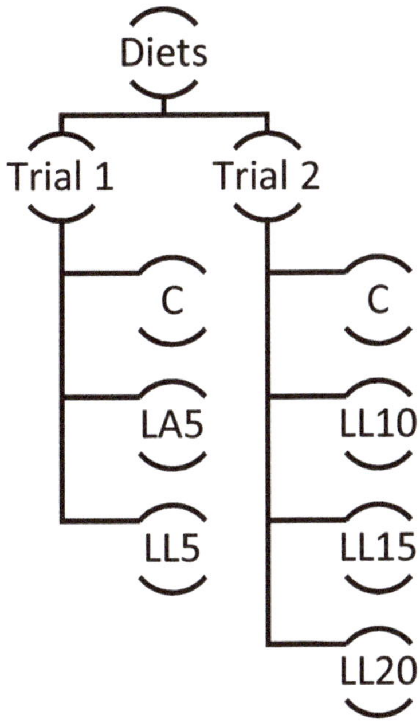

Figure 1. Diagram showing the distribution of diets by trial 1 and trial 2 with an indication of the level of the incorporation of *Lupinus albus* (LA) and Lupinus luteus (LL). C—control; LL5—*Lupinus luteus* 50 g/kg; LA5—*Lupinus albus* 50 g/kg; LL10—*Lupinus luteus* 100 g/kg; LL15—*Lupinus luteus* 150 g/kg; LL20—*Lupinus luteus* 200 g/kg; DM.

Details on diet formulation and chemical composition are found in Table 1, which is published by Almeida et al. [14].

Table 1. Chemical composition (g/kg DM) and levels of inclusion of diet components (g/kg as fed) of the different treatment diets.

Diets	Diet Components					Chemical Composition		
	Soybean Meal	Wheat	*Lupinus albus* (LA)	*Lupinus luteus* (LL)	Hay	DM	CP	NDF
Trial 1								
C	150	20	0	0	830	930	121	637
LL5	100	20	0	50	830	933	117	629
LA5	100	20	50	0	830	923	114	632
Trial 2								
C	170	20	0	0	810	929	130	626
LL10	100	20	0	100	780	925	127	621
LL15	50	20	0	150	780	937	128	628
LL20	0	20	0	200	780	941	128	628

C—control; LL5—*Lupinus luteus* 50 g/kg; LA5—*Lupinus albus* 50 g/kg; LL10—*Lupinus luteus* 100 g/kg; LL15—*Lupinus luteus* 150 g/kg; LL20—*Lupinus luteus* 200 g/kg; DM—dry matter; CP—crude-protein; NDF—neutral detergent fiber. Table adapted from Almeida et al. [14].

2.3. Carcass Traits, Cutting, and Dissection

Lambs were weighed weekly until they reached their target weight and then were submitted to a digestibility trial before slaughter, as described in Almeida et al. [14]. At the end of this trial, the animals were slaughtered at 23 ± 2 kg live weight using standard commercial procedures according to the Portuguese law on animal welfare in experimental research [15]. In trial 1, three animals per group were slaughtered. In trial 2, all animals were slaughtered except one from the control group that died at the end of the digestibility trial. Live weight at slaughter (LWS) was recorded after an overnight fast. Carcass dressing and dissection were performed after the methods of Fisher and DeBoer [16]. Carcasses were refrigerated for 24 h at 4 °C, and cold carcass weight (CCW) was recorded. The carcass dressing percentage was calculated on a CCW basis. The carcasses were then split along the vertebral column and the left side was divided into eight commercial cuts following the procedure of Santos et al. [17]. After weighing, each cut was dissected into muscle, bone, fat (subcutaneous and intermuscular fat), and residues (major blood vessels, ligaments, tendons, and thick connective tissue sheets associated with some muscles). Experienced operators performed the dissection in a controlled environment with room temperature below 20 °C.

2.4. Muscle Sampling

The Longissimus thoracis et lumborum muscle (LM) samples were collected 24 h postmortem between the 6th thoracic and 5th lumbar vertebrae from the right half of each carcass. The LM between the 6th and 12th thoracic vertebrae was frozen and stored at −20 °C for fatty acid analysis. The remaining portion of LM from the 12th thoracic vertebrae and 5th lumbar vertebrae was packaged in vacuum bags (Combivac, Felzmann, Linz, Austria) using a packaging machine (Minipack–Torre, SpA, MVS–35, Dalmine, Italy) and aged at 4 °C for 72 h. This portion was used for cooking losses and Warner–Bratzler shear force determinations.

2.5. Meat Quality Measurements

The pH was assessed at 1 h (pH_1) and 24 h (pH_{24}) postmortem in the LM, between 1st and 2nd lumbar vertebrae, using a pH meter equipped with a penetration electrode and thermometer (Hanna Instruments, HI–9025, Woonsocket, RI, USA). Meat color was measured on the LM surface after 60 min of blooming by placing the samples in containers covered with polyethylene film at 4 °C, using the L* (lightness), a* (redness), and b* (yellowness) color space [18] with a Minolta CR–10 colorimeter (Osaka, Japan). To assess cooking loss, LM samples of about 30 g were placed individually in polyethylene bags in a water bath at 78 °C. The samples were heated until an internal temperature of 75 °C was achieved and monitored with thermocouples. After being cooled for 15 min under running tap water, the samples were stored for 3 h at 4 °C, dried with filter paper, and weighed. Cooking loss was measured by comparing the final with the initial weight and is expressed as a percentage of the initial weight [19]. The meat samples used to determine the cooking loss were then cut into cuboid shape subsamples (3 to 4) of 1 cm^2 cross-section and 3–4 cm in length to determine the shear force, after room temperature equilibrium, using a Warner–Bratzler rectangular hole probe coupled to a Texture Analyser TA.XT plus texturometer with a load cell of 30 kgF (Stable Micro Systems, Godalming, UK). To perform this analysis, blade velocity and trigger force were set to 120 cm/min and 5 g, respectively, and the subsamples were placed with fibers perpendicular to the direction of the blade. Mean values for maximum shear force (kg/cm^2) over each subsample group were then obtained.

For fatty acid determination, the procedures described by Argemi-Armengol et al. [20] were followed. Briefly, the LM samples were trimmed of intermuscular and subcutaneous fat before fatty acid (FA) analysis. To determine fat content, the Ankom procedure (AOCS, 2005) [21] (Official Procedure Am 5–04) was applied using an Ankom extractor (XT10; Ankom Technology, Madrid, Spain). The FA methyl esters were obtained by transesterification using a 2% (*v*/*v*) methanol/sulfuric acid solution, with heating for 30 min at 80 °C, centrifugation at 3000 rpm for 5 min, and collection of the final supernatant. The FA methyl esters' analysis was performed in duplicate via gas chromatography with a 30 m × 0.25 mm capillary column and a flame ionization detector (Agilent DB-23; Agilent Technologies, Santa Clara, CA, USA). The helium was used as the carrier gas at a flow rate of two mL/min. The oven temperature was programmed to increase 35 °C per minute between 150 °C and 180 °C and at 5 °C per minute up to 220 °C. For the injector and detector, it was considered a temperature of 250 °C. The relative percentage of each FA in relation to the total FA was considered. The FAs were identified by comparing the retention times with a known standard Supelco® 37 Component FAME Mix (Supelco, Bellefonte, PA, USA). In total, 34 FAs were detected and quantified. The proportions of saturated fatty acids (SFA) (C10:0; C12:0; C13:0; C14:0; C15:0; C16:0; C17:0; C18:0; C20:0; C21:0; C22:0 and C23:0); polyunsaturated (PUFA) (C18: 2n − 6; C18: 3n − 3; C18: 3n − 6; C20: 2n − 6; C20: 3n − 6; C20: 3n − 3; C20: 4n − 6; C20: 5n − 3; C22: 5n − 3 and C22: 6n − 3); monounsaturated (MUFA) (C14: 1n − 5; C15: 1n − 5; C16: 1n − 7; C17: 1n − 7; C18: 1n − 9; C18: 1n − 7; C20: 1n − 9; C22: 1n − 9); cis/trans (9c,11t-C18:2; 9t,11t-C18:2; t11 C18:1; 9t-C18:1) were calculated.

2.6. Statistical Analyses

Data were subjected to an analysis of variance (ANOVA), performed on JMP®, version 14 [22]. Diets were used as the main factor. Tukey's multiple comparison test evaluated significant differences. Significance was declared at $p < 0.05$.

3. Results

3.1. Live Weight, Carcass Traits, and Meat Quality

In trial 1 and trial 2, no differences ($p > 0.05$) were found for LWS and CCW, although there was a tendency ($p = 0.06$) for dressing (CCW basis) % to be different in trial 1 (Table 2).

Table 2. Least square means (±SE) of live weight at slaughter, cold carcass weight, and carcass dressing of lambs in trials 1 and 2.

Trial	Diets	LWS (kg)	CCW (kg)	Dressing (CCW Basis) %
Trial 1	C (n = 3)	23.8 ± 0.82	9.18 ± 0.35	38.6 ± 1.22
	LA5 (n = 3	24.7 ± 1.56	9.3 ± 0.52	37.7 ± 0.83
	LL5 (n = 3)	23.7 ± 1.46	9.4 ± 0.43	40.0 ± 1.34
p		0.471	0.818	0.060
Trial 2	C (n = 3)	21.2 ± 1.96	8.38 ± 0.80	39.5 ± 2.26
	LL10 (n = 4)	21.2 ± 1.75	7.9 ± 0.89	37.0 ± 1.53
	LL15 (n = 3)	22.7 ± 0.54	8.6 ± 0.43	38.1 ± 1.44
	LL20 (n = 4)	22.1 ± 1.38	8.3 ± 0.98	37.5 ± 2.47
p		0.483	0.603	0.430

C—control; LL5—*Lupinus luteus* 50 g/kg; LA5—*Lupinus albus* 50 g/kg; LL10—*Lupinus luteus* 100 g/kg; LL15—*Lupinus luteus* 150 g/kg; LL20—*Lupinus luteus* 200 g/kg; LWS—live weight at slaughter; CCW—cold carcass weight.

For cuts percentage and cuts composition after dissection on both trials (data not shown), no differences were reported. For carcass composition in tissues and muscle:bone ratio, there were also no differences ($p > 0.05$) among diets for either trial 1 or trial 2 (Table 3).

Table 3. Least square means (±SE) of carcass composition in tissues (%) and muscle:bone ratio of lambs in trials 1 and 2.

Trial	Diets	Carcass Tissues (%)				Muscle:Bone
		Muscle	Bone	Intermuscular Fat	Subcutaneous Fat	
Trial 1	C (n = 3)	58.0 ± 1.30	26.1 ± 0.74	11.7 ± 1.37	2.9 ± 0.12	2.0 ± 0.40
	LA5 (n = 3)	58.3 ± 1.92	25.7 ± 1.39	11.0 ± 0.15	3.2 ± 0.76	1.7 ± 0.32
	LL5 (n = 3)	58.2 ± 1.13	25.8 ± 0.77	11.8 ± 0.85	2.7 ± 0.21	2.1 ± 0.36
p		0.977	0.870	0.466	0.441	0.366
Trial 2	C (n = 3)	56.0 ± 1.64	31.4 ± 1.42	8.3 ± 2.79	3.8 ± 0.83	1.5 ± 0.15
	LL10 (n = 4)	55.7 ± 2.22	31.4 ± 1.79	7.2 ± 1.99	4.9 ± 0.61	1.3 ± 0.12
	LL15 (n = 4)	57.0 ± 1.22	30.2 ± 1.67	7.9 ± 1.86	4.5 ± 0.60	1.9 ± 0.14
	LL20 (n = 4)	53.6 ± 2.77	31.8 ± 2.72	9.1 ± 2.54	4.8 ± 0.93	1.4 ± 0.28
p		0.191	0.690	0.710	0.238	0.339

C—control; LL5—*Lupinus luteus* 50 g/kg; LA5—*Lupinus albus* 50 g/kg; LL10—*Lupinus luteus* 100 g/kg; LL15—*Lupinus luteus* 150 g/kg; LL20—*Lupinus luteus* 200 g/kg.

No effect ($p > 0.05$) of diet on pH, L*a*b* color parameters, and cooking loss was observed. Lambs fed only *Lupinus luteus* (LL20) presented higher shear force values than those fed LL10, hence tougher meat ($p = 0.045$; Table 4). However, the shear force was similar between lamb meat and the rest of the groups. Although pH was not different ($p > 0.05$) between diets in both trials, it is important to point out that pH values were particularly high in the second trial, specially pH1.

Table 4. Least square means (±SE) of meat quality attributes of lambs in trials 1 and 2.

Trial	Diets	pH		Color			CL (%)	SF (kg/cm^2)
		pH$_1$	pH$_{24}$	L*	a*	b*		
Trial 1	C (n = 3)	6.7 ± 0.06	6.0 ± 0.33	47.1 ± 3.64	16.8 ± 0.93	15.8 ± 0.63	11.6 ± 1.16	4.9 ± 0.40
	LA5 (n = 3)	6.7 ± 0.07	5.8 ± 0.08	46.3 ± 2.13	14.7 ± 1.77	15.0 ± 0.80	11.8 ± 2.38	5.3 ± 0.79
	LL5 (n = 3)	6.7 ± 0.04	5.8 ± 0.10	46.0 ± 2.71	15.5 ± 0.57	15.4 ± 1.06	11.2 ± 2.03	5.2 ± 0.54
p		0.759	0.147	0.876	0.157	0.388	0.934	0.739
Trial 2	C (n = 3)	6.8 ± 0.63	6.6 ± 0.36	45.3 ± 2.73	16.1 ± 0.24	13.7 ± 0.77	8.5 ± 0.94	3.9 ± 1.18 ^{ab}
	LL10 (n = 4)	7.4 ± 0.27	6.4 ± 0.15	46.4 ± 4.10	14.3 ± 2.15	13.9 ± 1.10	9.1 ± 1.48	3.8 ± 0.89 ^{b}
	LL15 (n = 4)	7.8 ± 0.49	6.4 ± 0.11	45.8 ± 1.60	16.3 ± 2.20	14.7 ± 0.78	9.4 ± 2.59	4.2 ± 0.92 ^{ab}
	LL20 (n = 4)	7.3 ± 0.42	6.5 ± 0.43	44.9 ± 3.35	15.1 ± 1.98	12.8 ± 1.32	8.7 ± 1.49	4.7 ± 1.09 ^{a}
p		0.477	0.860	0.915	0.475	0.144	0.728	0.045

C—control; LL5—*Lupinus luteus* 50 g/kg; LA5—*Lupinus albus* 50 g/kg; LL10—*Lupinus luteus* 100 g/kg; LL15—*Lupinus luteus* 150 g/kg; LL20—*Lupinus luteus* 200 g/kg; CL—cooking losses; SF—shear force. Different superscript letters (a, b) on the same column indicate significant differences ($p < 0.05$).

3.2. Fatty Acids

Table 5 shows fatty acids with values greater than 1% (g/100 g identified FA). For both trials, the fat composition comprises SFA, MUFA, and PUFA fatty acids, with palmitic (C16:0), stearic (C18:0), and oleic (c9 C18:1) fatty acids being the most abundant. Among these FAs, oleic acid (c9 C18:1) showed the highest percentage, ranging from 29% to 36%, followed by palmitic acid (C16:0) (from 20% to 23%) and stearic (C18:0) (from 18% to 23%). In general, diets did not have a significant effect on FAs. However, in trial 1, diets show a significant effect ($p = 0.002$) on t11 C18:1, in which the LL5 diet showed higher values than the LA5 diet and was similar to the control. A similar trend was observed for C16:0. It was also observed that the LM of lambs on the LL5 and LA5 diets showed higher values of oleic MUFAs ($p < 0.05$) than the control (Table 5).

Table 5. Fatty acid (FA) composition (g/100 g identified FA) in LM of lambs affected by diets in trials 1 and 2.

Trial	Diets	SFA				MUFA					PUFA	
		C14:0	C16:0	C17:0	C18:0	C15:1n−5	C16:1n−7	C18:1n−7	t11 C18:1	c9 C18:1	C18:2n−6	C20:4n−6
Trial 1	C (n = 3)	3.81	22.7 [a]	1.56	17.86	1.57	1.25	1.14	1.78 [a]	32.3 [b]	6.42	2.76
	LA5 (n = 3)	4.02	21.6 [b]	1.38	17.17	1.31	1.23	1.02	1.40 [b]	35.7 [a]	6.28	2.42
	LL5 (n = 3)	3.19	22.4 [a]	1.42	18.24	1.36	1.29	0.98	1.86 [a]	34.2 [a]	5.95	2.74
p		0.577	0.019	0.554	0.837	0.588	0.960	0.772	0.002	0.013	0.839	0.707
Trial 2	C (n = 3)	3.29	19.7	1.40	18.1	2.70	0.98	1.26	1.73	29.1	9.27	5.38
	LL10 (n = 4)	2.24	20.4	1.52	23.1	1.67	0.89	0.92	1.48	30.8	7.01	3.70
	LL15 (n = 4)	2.93	20.3	1.33	20.0	1.96	1.02	0.97	1.64	31.3	8.14	3.97
	LL20 (n = 4)	3.17	19.5	1.21	20.6	2.06	1.08	0.93	1.55	30.0	7.96	4.97
p		0.512	0.898	0.208	0.268	0.222	0.824	0.064	0.806	0.750	0.126	0.496

C—control; LA5—*Lupinus albus* 50 g/kg; LL5—*Lupinus luteus* 50 g/kg; LL10—*Lupinus luteus* 100 g/kg; LL15—*Lupinus luteus* 150 g/kg; LL20—*Lupinus luteus* 200 g/kg; FA—fatty acid; SFA—saturated FA; MUFA—monounsaturated FA; PUFA—polyunsaturated FA. Different superscript letters (a, b) on the same column indicate significant differences ($p < 0.05$).

No differences were found among diets in most FA groups, except for MUFA in trial 1. In this case, diets with lupin inclusion had a higher content ($p < 0.05$) of MUFA fatty acids (Table 6). Regardless, similarly to what was observed for individual FA in all alternative sources of protein diets, the results were very similar to those of the control diet.

Table 6. Groups of SFA, MUFA and PUFA (g/100 g identified FA) in LM of lambs affected by diets in trials 1 and 2.

Trial	Diets	SFA	MUFA	PUFA	n-3	n-6	n-6/n-3
Trial 1	C (n = 3)	47.6	39.4 [b]	12.8	2.18	10.1	4.73
	LA5 (n = 3)	45.8	42.2 [a]	12.1	2.04	9.45	4.64
	LL5 (n = 3)	46.7	41.2 [a]	12.1	2.01	9.48	4.72
p		0.492	0.031	0.776	0.541	0.865	0.985
Trial 2	C (n = 3)	44.0	37.2	18.6	2.34	15.75	6.74
	LL10 (n = 4)	48.8	37.1	14.1	1.95	11.69	6.08
	LL15 (n = 4)	46.1	38.2	15.5	1.99	13.04	6.67
	LL20 (n = 4)	46.2	37.1	16.6	2.00	14.05	7.08
p		0.200	0.943	0.353	0.781	0.357	0.093

C—control; LA5—*Lupinus albus* 50 g/kg; LL5—*Lupinus luteus* 50 g/kg; LL10—*Lupinus luteus* 100 g/kg; LL15—*Lupinus luteus* 150 g/kg; LL20—*Lupinus luteus* 200 g/kg; FA—fatty acids; SFA—saturated FA; MUFA—monounsaturated FA; PUFA—polyunsaturated FA; LC—long chain. Different superscript letters (a, b) on the same column indicate significant differences ($p < 0.05$).

4. Discussion

Lupin incorporation had no overall effect on carcass characteristics, although dressing percentages of all the carcasses were lower than the ones observed in younger and lighter lambs of the same breed [17,23], as well as on lambs with similar LWS [9,10]. However, it is important to note that lambs in the studies of Santos et al. [23] and Santos et al. [17] were slaughtered around 8–11 kg LW (around weaning) and produced in accordance with the Protected Denomination of 89 Origin (PDO) specifications for carcasses of "Borrego terrincho–PDO", which might explain the differences in dressing percentages. Dietary lupin incorporation also had no influence ($p > 0.05$) on the percentage of cuts in the carcass, which was reasonably similar to the values observed by other authors [9,10,24]. Overall, muscle content in the carcass was lower than the content previously observed in carcasses of lambs of the same breed, considering the values reported by Santos et al. [17], which, in turn, led to lower muscle:bone ratios than the ratios observed by these authors. This can be explained by the fact that the animals studied by Santos et al. [17] were suckling lambs. Since there were no statistical differences ($p > 0.05$) among groups, this could indicate that the inclusion of lupins in lambs' diets will not impact the dressing percentages, although this could also be due to the small sample size in this study. Muscle:bone ratios were lower than expected for this breed [17]. Leg muscle:bone ratio, for example, was slightly lower than the 2.64 value reported by Lestingi et al. [10] when incorporating 23% of *Lupinus albus* in Gentile di Puglia lambs' diets. However, the same study reported lower loin muscle:bone values (1.54), due to a far higher percentage of bone than the one found in this study, which might be attributed to phenotypic differences between the two breeds.

Particularly in trial 2, muscle pH_{24} was higher than desirable [17] and reported by other authors using lupins in lambs' diets [25,26]. During the preslaughter period, all lambs were handled by the same person to reduce acute stress as much as possible since it can increase ultimate meat pH [26]. Despite this outcome, there seems to have been no effect of diet on this parameter ($p > 0.05$). Meat color was within the values previously found for this breed [17]. Other authors have reported darker meats than the ones in this study, in crossbred lambs and Merino wether weaner sheep fed diets with 20–35% (DM) lupin incorporation [27,28] ($L^* = 37.5 \pm 0.8$; $L^* = 39.0 \pm 0.6$, respectively). The values of b* of LL20 lambs ($b^* = 12.1$), which were fed the highest lupin inclusion of both trials, were similar to the ones reported by White et al. [28] with higher lupin inclusions ($b^* = 12.8$). In the present study, lupin inclusion in the diets resulted in slightly tougher lamb meat (4.70 kg/cm^2; $p = 0.045$), as was previously reported [29] for lambs fed a mixture of barley and lupins. Although Santos et al. [17] reported far lower shear force values in lamb meat from the same breed, the methodology applied by these authors was different and, therefore, not comparable to the values found in this study. Other authors have reported lower shear force values than those in this study [9,24] (2.48 and 2.04 kg/cm^2, respectively); however, overall, lupin inclusion in lambs' diets tends not to have an impact on meat tenderness. Higher sample size would probably clarify the results and provide a clearer conclusion than these preliminary results. Since higher meat pH can be associated with higher tenderness [25], this might also explain the present study results. While no differences were found in either trial for cooking losses ($p > 0.05$), the high meat pH values found in the second trial might also justify the lower cooking losses [30] observed in these lambs' meat (9% vs. 11% in trial 1).

The evaluation of the fatty acid composition of the LM of lambs fed diets containing alternative protein sources muscle has been the target of several researchers [9,31]. Diets including faba beans (*Vicia faba*), peas (*Pisum sativum*), and lupins (*Lupinus* sp.) to replace soybean meal have been tested over the years to understand the effects on the intramuscular FAs of lambs [9,29,32,33]. In the present study, the most abundant fatty acid in the intramuscular fat of lambs was oleic acid (c9 C18:1), which has been positively associated with human health [34,35] and did not differ among treatments except for trial 1, where the incorporation of Lupins proved to have a positive effect on this FA. These results are consistent with those of other studies in which alternative protein sources were used [29,36]. However, reports from other studies testing alternative protein sources show variable effects on this FA. For example, Lestingi et al. [9] found that the meat of lambs fed lupins contained lower levels of oleic acid (c9 C18:1) than peas or the combination of lupin and peas, and Lanza et al. [32] reported that oleic acid in intramuscular fat was higher ($p < 0.05$) in the pea group that that in the soybean-meal group. On the other hand, vaccenic acid (t11 C18:1) showed a higher value ($p < 0.05$) in trial 1 for the group in which *Lupinus albus* was used but had no effect when compared with the groups with *Lupinus luteus* and soybean meal. Differences in vaccenic acid were also verified by other authors who studied alternative legume seeds as a protein alternative to soybean meal [32,37]. The reason for this is unclear, but differences in the fatty acid composition may depend on incomplete ruminal hydrogenation of dietary fat [37].

Unlike the results found in this study, other authors showed differences in PUFAs. For example, Scerra et al. [29] compared diets containing pea, faba bean, and soybean meal and concluded that the meat of lambs fed with peas had higher proportions of the essential fatty acids C18: $2n - 6$ and C18: $3n - 3$ than those of the other groups. On the other hand, faba beans led to higher proportions of PUFAs ($p < 0.01$) in lamb meat than sweet lupins and peas, according to Gómez-Cortés et al. [33]. Regarding the $n - 6/n - 3$ fatty acid ratio, the nutritional guidelines for human consumption recommend optimizing the intake of foods containing high amounts of n − 3 fatty acids, which is proven to reduce the incidence of cardiovascular diseases [38]. Our results show that the n − 6/n − 3 fatty acid ratio is similar in meats from lambs fed on lupin and soybean meal diets. Additionally, the values for this ratio are similar to those reported by other authors who tested lupins and other legume grains [9,32].

5. Conclusions

Lupin incorporation in diets of growing lambs produced similar results to soybean meal in terms of carcass traits and meat physical characteristics. Replacement of soybean meal by lupins in fattening lambs' diets had minor effects on their intramuscular FA profile. However, adding legume grains such as lupins to lamb diets seemed to have no effect on the n-6/n-3 ratio. Nevertheless, the lack of significant differences could also be attributed to the small sample size. Combined with the results observed during the growth and digestibility trials [14], lupin incorporation in lambs' diets can present a solution for some producers. Further research should be conducted with a larger sample to provide more information and study the effect of lupin inclusion on the palatability of lambs' meat.

Author Contributions: Conceptualization, C.G., L.F., S.S., J.A. and V.S.; methodology, C.G., L.F., S.S., V.S., M.A. and S.G.-S.; software, M.A., S.S. and V.S.; investigation, M.A., D.C., A.A., S.G.-S., S.S., V.S., C.G., J.A.S. and J.M.L.; writing—original draft preparation, M.A.; writing—review and editing, M.A., S.G.-S., D.C., A.A., J.A.S., J.M.L., J.A., V.S., C.G., S.S. and L.F.; supervision, L.F., C.G., S.S. and V.S.; project administration, C.G.; funding acquisition, C.G. All authors have read and agreed to the published version of the manuscript.

Funding: The authors acknowledge financial support through project INTERACT–"Integrative Research in Environment, Agro-Chains, and Technology", reference NORTE-01-0145-FEDER-000017, in its line of research entitled ISAC, co-financed by the European Regional Development Fund through NORTE 2020 (North Regional Operational Program 2014/2020).

Institutional Review Board Statement: This study was conducted according to the guidelines of the Portuguese law on animal welfare in experimental research, and the protocol was approved by the ORBEA (Animal Welfare Body) of the University Trás-os-Montes and Alto Douro (669-e-DZ-2018).

Data Availability Statement: The data presented in this study is available on request.

Acknowledgments: Authors integrated into the research unit CECAV and All4AnimalS were supported by the projects UIDB/CVT/00772/2020 and LA/P/0059/2020 funded by the Portuguese Foundation for Science and Technology (FCT). Jose M. Lorenzo is a member of the HealthyMeat network, funded by CYTED (ref. 119RT0568). The authors thank and acknowledge the University of Trás-os-Montes and Alto Douro experimental farm and animal nutrition laboratory staff.

Conflicts of Interest: The authors declare no conflict of interest.

References

1. Watson, C.A.; Reckling, M.; Preissel, S.; Bachinger, J.; Bergkvist, G.; Kuhlman, T.; Lindström, K.; Nemecek, T.; Topp, C.; Vanhatalo, A.; et al. Grain legume production and use in European agricultural systems. *Adv. Agron.* **2017**, *144*, 235–303. [CrossRef]
2. Gresta, F.; Wink, M.; Prins, U.; Abberton, M.; Capraro, J.; Scarafoni, A.; Hill, G. Lupins in European cropping systems. In *Legumes in Cropping Systems*; Murphy-Bokern, D., Stoddard, F.L., Watson, C.A., Eds.; CABI: Surrey, UK, 2017; pp. 88–108. [CrossRef]
3. Day, L. Proteins from land plants–potential resources for human nutrition and food security. *Trends Food Sci. Technol.* **2013**, *32*, 25–42. [CrossRef]
4. van Barneveld, R.J. Understanding the nutritional chemistry of lupin (*Lupinus* spp.) seed to improve livestock production efficiency. *Nutr. Res. Rev.* **1999**, *12*, 203–230. [CrossRef]
5. Martín-Pedrosa, M.M.; Varela, A.; Guillamon, E.; Cabellos, B.; Burbano, C.; Comez-Fernández, J.; de Mercado, E.; Gomez-Izquierdo, E.; Cuadrado, C.; Muzquiz, M. Biochemical characterization of legume seeds as ingredients in animal feed. *Span. J. Agric. Res.* **2016**, *14*, 16. [CrossRef]
6. Masucci, F.; Di Francia, A.; Romano, R.; di Serracapriola, M.M.; Lambiase, G.; Varricchio, M.L.; Proto, V. Effect of *Lupinus albus* as protein supplement on yield, constituents, clotting properties and fatty acid composition in ewes' milk. *Small Rumin. Res.* **2006**, *65*, 251–259. [CrossRef]
7. Ata, M.; Obeidat, B.S. The inclusion of sweet lupin grain (*Lupinus angustifolius*) improves nursing performance of lactation in Awassi ewes. *Small Rumin. Res.* **2020**, *190*, 106150. [CrossRef]

8. Facciolongo, A.M.; Rubino, G.; Zarrilli, A.; Vicenti, A.; Ragni, M.; Toteda, F. Alternative protein sources in lamb feeding 1. Effects on productive performances, carcass characteristics and energy and protein metabolism. *Prog. Nutr.* **2014**, *16*, 105–115.
9. Lestingi, A.; Facciolongo, A.M.; Jambrenghi, A.C.; Ragni, M.; Toteda, F. The use of peas and sweet lupin seeds alone or in association for fattening lambs: Effects on performance, blood parameters and meat quality. *Small Rumin. Res.* **2016**, *143*, 15–23. [CrossRef]
10. Lestingi, A.; Colonna, M.A.; Marsico, G.; Tarricone, S.; Facciolongo, A.M. Effects of legume seeds and processing treatment on growth, carcass traits and blood constituents of fattening lambs. *S. Afr. J. Anim. Sci.* **2019**, *49*, 799–809. [CrossRef]
11. Fychan, R.; Marley, C.; Lewis, G.; Davies, R.; Theobald, V.; Jones, R.; Abberton, M. Effects of feeding concentrate diets containing narrow-leafed lupin, yellow lupin or soya when compared with a control diet on the productivity of finishing lambs. In Proceedings of the 12th International Lupin Conference—Lupins for Health and Wealth, Fremantle, Australia, 14–18 September 2008; pp. 127–130, ISBN 0-86476-153-8.
12. Food and Agriculture Organization of the United Nations (FAOSTAT). *Food Supply-Livestock and Fish Primary Equivalent*; FAOSTAT: Rome, Italy, 2018. Available online: http://www.fao.org/faostat/en/#data/QL (accessed on 9 October 2020).
13. de Rancourt, M.; Fois, N.; Lavín, M.P.; Tchakérian, E.; Vallerand, F. Mediterranean sheep and goats production: An uncertain future. *Small Rumin. Res.* **2006**, *62*, 167–179. [CrossRef]
14. Almeida, M.; Garcia-Santos, S.; Nunes, A.; Rito, S.; Azevedo, J.; Guedes, C.; Silva, S.; Ferreira, L. Introducing Mediterranean Lupins in Lambs' Diets: Effects on Growth and Digestibility. *Animals* **2021**, *11*, 942. [CrossRef] [PubMed]
15. Decree-Law No. 1/2019 Amending Decree-Law No. 113/2013 Implementing EU Directive No. 2010/63 on Animal Protection for Scientific Purposes, Government of Portugal. Available online: http://www.fao.org/faolex/results/details/en/c/LEX-FAOC183382 (accessed on 30 June 2022).
16. Fisher, A.V.; de Boer, H. The EAAP standard method of sheep carcass assessment. Carcass measurements and dissection procedures. Report of the EAAP Working Group on Carcass Evaluation, in cooperation with the CIHEAM Instituto Agronomico Mediterraneo of Zaragoza and the CEC Directorate General for Agriculture in Brussels. *Livest. Prod. Sci.* **1994**, *38*, 149–156.
17. Santos, V.A.C.; Silva, S.R.; Azevedo, J.M.T. Carcass composition and meat quality of equally mature kids and lambs. *J. Anim. Sci.* **2008**, *86*, 1943–1950. [CrossRef] [PubMed]
18. Commission International de l'Eclairage (CIE). *Colorimetry*, 2nd ed.; CIE Publications: Vienna, Austria, 1986; Volume 15.2.
19. Honikel, K.O. How to measure the water holding capacity of meat? Recommendation of standardized method. In *Evaluation and Control of Meat Quality in Pigs*; Tarrant, P.V., Eikelenboom, G., Monin, G., Eds.; Martinus Nijhoff: Leiden, The Netherlands, 1987; pp. 129–142.
20. Argemí-Armengol, I.; Villalba, D.; Tor, M.; Pérez-Santaescolástica, C.; Purriños, L.; Lorenzo, J.M.; Álvarez-Rodríguez, J. The extent to which genetics and lean grade affect fatty acid profiles and volatile compounds in organic pork. *PeerJ* **2019**, *7*, e7322. [CrossRef] [PubMed]
21. AOCS, Official Method Am 5-04. Rapid determination of oil/fat utilizing high-temperature solvent extraction. In *Official Methods and Recommended Practices of the AOCS*, 6th ed.; Firestone, D., Ed.; AOCS Press: Urbana, IL, USA, 2013.
22. SAS Institute Inc. *JMP® 14*; SAS Institute Inc.: Cary, NC, USA, 2018.
23. Santos, V.A.C.; Silva, S.R.; Mena, E.G.; Azevedo, J.M.T. Live weight and sex effects on carcass and meat quality of "Borrego terrincho–PDO" suckling lambs. *Meat Sci.* **2007**, *77*, 654–661. [CrossRef] [PubMed]
24. Facciolongo, A.M.; De Marzo, D.; Ragni, M.; Lestingi, A.; Toteda, F. Use of alternative protein sources for finishing lambs. 2. Effects on chemical and physical characteristics and fatty acid composition of meat. *Prog. Nutr.* **2015**, *17*, 165–173.
25. Wiese, S.C.; White, C.L.; Masters, D.G.; Milton, J.T.B.; Davidson, R.H. Growth and carcass characteristics of prime lambs fed diets containing urea, lupins or canola meal as a crude protein source. *Aust. J. Exp. Agric.* **2003**, *43*, 1193–1197. [CrossRef]
26. White, C.L.; Hanbury, C.D.; Young, P.; Phillips, N.; Wiese, S.C.; Milton, J.B.; Davidson, R.H.; Siddique, K.H.M.; Harris, D. The nutritional value of *Lathyrus cicera* and *Lupinus angustifolius* grain for sheep. *Anim. Feed Sci. Technol.* **2002**, *99*, 45–64. [CrossRef]
27. Warner, R.D.; Ferguson, D.M.; McDonagh, M.B.; Channon, H.A.; Cottrell, J.J.; Dunshea, F.R. Acute exercise stress and electrical stimulation influence the consumer perception of sheep meat eating quality and objective quality traits. *Aust. J. Exp. Agric.* **2005**, *45*, 553–560. [CrossRef]
28. Ponnampalam, E.N.; Sinclair, A.J.; Hosking, B.J.; Egan, A.R. Effects of dietary lipid type on muscle fatty acid composition, carcass leanness, and meat toughness in lambs. *J. Anim. Sci.* **2002**, *80*, 628–636. [CrossRef]
29. Scerra, M.; Caparra, P.; Foti, F.; Cilione, C.; Zappia, G.; Motta, C.; Scerra, V. Intramuscular fatty acid composition of lambs fed diets containing alternative protein sources. *Meat Sci.* **2011**, *87*, 229–233. [CrossRef] [PubMed]
30. Bouton, P.E.; Harris, P.T.; Shorthose, W.R. Effect of ultimate pH upon the water-holding capacity and tenderness of mutton. *J. Food Sci.* **1971**, *36*, 435–439. [CrossRef]
31. Halmemies-Beauchet-Filleau, A.; Rinne, M.; Lamminen, M.; Mapato, C.; Ampapon, T.; Wanapat, M.; Vanhatalo, A. Alternative and novel feeds for ruminants: Nutritive value, product quality and environmental aspects. *Animal* **2018**, *12* (Suppl. S2), s295–s309. [CrossRef] [PubMed]
32. Lanza, M.; Fabro, C.; Scerra, M.; Bella, M.; Pagano, R.; Brogna, D.M.R.; Pennisi, P. Lamb meat quality and intramuscular fatty acid composition as affected by concentrates including different legume seeds. *Ital. J. Anim. Sci.* **2011**, *10*, e18. [CrossRef]
33. Gómez-Cortés, P.; Galisteo, O.O.; Ramírez, C.A.; Blanco, F.P.; de la Fuente, M.A.; Sánchez, N.N.; Marín, A.L.M. Intramuscular fatty acid profile of feedlot lambs fed concentrates with alternative ingredients. *Anim. Prod. Sci.* **2019**, *59*, 914–920. [CrossRef]

34. Sales-Campos, H.; de Souza, P.R.; Peghini, B.C.; da Silva, J.S.; Cardoso, C.R. An overview of the modulatory effects of oleic acid in health and disease. *Mini Rev. Med. Chem.* **2013**, *13*, 201–210.
35. Martins, T.; de Lemos, M.V.; Mueller, L.F.; Baldi, F.; de Amorim, T.Y.; Ferinho, A.M.; Muñoz, J.A.; Fuzikawa, I.; Moura, G.; Gemelli, J.; et al. Fat deposition, fatty acid composition, and its relationship with meat quality and human health. *Meat Sci. Nutr.* **2018**, *78*, 17–37.
36. Lanza, M.; Bella, M.; Priolo, A.; Fasone, V. Peas (*Pisum sativum* L.) as an alternative protein source in lamb diets: Growth performances, and carcass and meat quality. *Small Rumin. Res.* **2003**, *47*, 63–68. [CrossRef]
37. Bonanno, A.; Tornambè, G.; Di Grigoli, A.; Genna, V.; Bellina, V.; Di Miceli, G.; Giambalvo, D. Effect of legume grains as a source of dietary protein on the quality of organic lamb meat. *J. Sci. Food Agric.* **2012**, *92*, 2870–2875. [CrossRef]
38. EFSA, Panel on Dietetic Products, Nutrition, and Allergies (NDA). Scientific Opinion on Dietary Reference Values for fats, including saturated fatty acids, polyunsaturated fatty acids, monounsaturated fatty acids, trans fatty acids, and cholesterol. *EFSA J.* **2010**, *8*, 1461. [CrossRef]

animals

MDPI

Article

Effect of Dietary Inclusion of *Azadirachta indica* and *Moringa oleifera* Leaf Extracts on the Carcass Quality and Fatty Acid Composition of Lambs Fed High Forage Total Mixed Rations

Edward Cottington Webb, Abubeker Hassen *, Michael Olanrewaju Olaniyi and Pamela Pophiwa

Department of Animal Science, University of Pretoria, Hatfield, Pretoria 0028, South Africa
* Correspondence: abubeker.hassen@up.ac.za; Tel.: +27-12-420-3273

Simple Summary: Feed additives based on medicinal plants, such as neem and moringa plant extracts, are used to mitigate rumen methane emissions, but data regarding their effects on lamb meat quality are scarce. This study investigated the effects of oral supplementation of neem and moringa leaf extracts on the carcass quality and meat fatty acid composition of lambs. Neem leaf extracts had no effect on carcass fat and meat fatty acid composition. Whereas, Moringa leaf extract improved the meat fatty acid composition of lambs compared to the monensin treatment.

Abstract: There is an increased interest in the use of medicinal plants as alternatives to antibiotic growth promoters and as agents for methane production mitigation. This study investigated the effects of *Azadirachta indica* and *Moringa oleifera* feed additives on the carcass and meat quality of lambs. Forty South African Mutton Merino lambs, weighing between 29 and 43 kg, were randomly assigned to four treatment groups (n = 10 lambs/treatment) and fed a basal total mixed ration (TMR) containing soybean meal (17%), yellow maize (28%), Alfalfa hay (20%), Eragrostis curvula hay (22.2%), molasses (6.0%), wheat offal (5%), urea (0.8%) and vitamin premix (0.5%) on a DM basis. The dietary treatments: TMR diet (control); TMR diet with *A. indica* leaf extract (*A. indica* leaf extract at a dosage of 50 mg per kg of feed: neem); TMR diet with *M. oleifera* leaf extract (*M. oleifera* leaf extract at a dosage of 50 mg per kg DM of feed: moringa); TMR diet with monensin (at a dosage of 50 mg monensin sodium per kg of feed: positive control). After an adaptation period of 10 days to the experimental conditions, the lambs from all treatment groups were fed ad libitum with the experimental diets. The lambs were slaughtered at a live weight of 60–65 kg after a 23 week trial period. The plant extract dietary additives had no significant effects on the carcass characteristics of the lambs. In comparison to monensin, supplementing with moringa leaf extracts resulted in a higher proportion of C18:1n9c (45.0% ± 0.57 vs. 40.5% ± 0.80; $p < 0.05$), total MUFAs (47.3% ± 0.66 vs. 42.6% ± 0.87; $p < 0.05$), and UFA:SFA ratio (1.01 ± 0.03 vs. 0.85 ± 0.03; $p < 0.05$), which may be beneficial for human health. Our results suggest that natural feed additives, such as *A. indica* and *M. oleifera* leaf extracts, can be included in lamb diets without compromising meat fatty acid composition. The negative economic impacts of such technologies on animal production and farm profitability should not be expected.

Keywords: bioactive compounds; methane mitigation; carotenoids; phytochemicals; medicinal plants; feed additives

Citation: Webb, E.C.; Hassen, A.; Olaniyi, M.O.; Pophiwa, P. Effect of Dietary Inclusion of *Azadirachta indica* and *Moringa oleifera* Leaf Extracts on the Carcass Quality and Fatty Acid Composition of Lambs Fed High Forage Total Mixed Rations. *Animals* **2022**, *12*, 2039. https://doi.org/10.3390/ani12162039

Academic Editors: Panagiotis Simitzis and Athanasios I. Gelasakis

Received: 8 June 2022
Accepted: 5 August 2022
Published: 11 August 2022

1. Introduction

There are increasing concerns regarding the long-term use of antibiotics in animal feed due to the fact of their potential impact on the environment and human health [1–4]. This fact has generated increased interest in research on the use of medicinal plants as a safe and inexpensive approach to replace the use of antibiotic growth promoters [5]. Multipurpose plants, such as *Azadirachta indica* and *Moringa oleifera*, are used in animal production for their medicinal and antioxidant properties and as possible mitigants for rumen methane emission [6]. The methane inhibitory effect of these plants is related to the presence of bioactive compounds, such as alkaloids, flavonoids and tannins, which are capable of interacting with rumen microbes and influencing ruminal fermentation patterns [6,7].

Although dietary inclusion of *A. indica* and *M. oleifera* and other plant extracts are promising strategies for mitigating rumen methane production [6], feeding strategies that alter rumen fermentation patterns may also affect lipid metabolism and, subsequently, meat quality [7]. Research has shown that phytogenic feed additives have an antimicrobial effect on bacterial species involved in rumen biohydrogenation, resulting in improved absorption and accumulation of polyunsaturated fatty acids (PUFAs) and conjugated linoleic acids (CLAs) in milk [8] and presumably meat. A previous study showed that supplementation with up to 5% neem fruit in lamb diets was effective in achieving high concentrations of rumenic acid (C18:2 cis-9 trans-11), a CLA which is beneficial for human health [9]. Feeding moringa silage (rich in α-linolenic acid) has been reported to increase both n-3 PUFA and CLA in lamb meat [10].

Most studies on *A. indica* and *M. oleifera* supplementation in ruminant diets focus on feeding whole plant parts, such as leaves, pods, fruits and seeds, to animals [9–11]. Plant extracts are often used in ethnoveterinary medicine and their inclusion as dietary additives is a useful strategy to conserve medicinal plants [12,13]. In vitro studies have shown the antimethanogenic properties of *A. indica* and *M. oleifera* plant extracts [14]; however, the effect of such antimethanogenic feed additives on animal production and meat quality has not been thoroughly examined. Farmers are less likely to adopt new technologies unless they are cost effective and induce no negative effects on animal production and product quality. Therefore, the objective of the current research was to investigate the effects of *A. indica* and *M. oleifera* plant extracts on the carcass traits and meat fatty acid composition of South African Mutton Merino (SAMM) lambs.

2. Materials and Methods

2.1. Collection of Plant Materials and Extraction Procedure

The leaves from *A. indica* and *M. oleifera* trees were harvested in the South West region of Nigeria. The harvesting and handling of the plant materials from Nigeria to the University of Pretoria, South Africa, have previously been described [14]. On arrival, the leaves were freeze-dried for 5 days to a constant weight and milled through a 0.5 mm screen. The ground samples (100 g) were extracted with methanol (1 L) in glass vials and placed in a shaker for 96 h. The extracts were sieved through a 150 μm screen aperture, precipitated, freeze-dried to a constant weight and stored at 4 °C.

2.2. Management of Experimental Animals

Forty weaned ram lambs (approximately 120–135 days old) with a mean body weight of 38.1 kg ± 3.83 were used in the present study in a completely randomised block design. The lambs were randomly allocated to four dietary treatment groups. Two lambs from each treatment within a block were allotted in a covered pen, with five pens of two lambs per treatment and a total of ten lambs per treatment. Each pen measured 3.2 × 2.2 m. The pens were considered experimental units, and the two sheep in each pen were the observational units. The lambs were kept at the Hatfield Experimental Farm of the University of Pretoria, in the city of Pretoria, South Africa.

A total mixed ration (TMR) was formulated by a commercial feed company to meet the growth and maintenance requirements of the lambs. The TMR was sampled to conduct a chemical analysis. The dry matter (DM) and ash of the TMR used in this study were determined according to the standard procedure described by the Association of Official Analytical Chemists (AOAC) [15]. Acid detergent fibre (ADF), neutral detergent fibre (NDF) and lignin were determined using an Ankom technology 200/220 (Ankom Technology, Fairport, NY, USA) as described by Van Soest et al. [16]. Nitrogen was analysed using a Leco Instrumente GmbH, Kirchheim, Germany, nitrogen/protein analyser. The ether extract was determined by extracting the sample with ether following the Tecator Soxtec

(HT6) system [15]. The formulation and chemical composition of the TMR is presented in Table 1.

Table 1. Formulation and chemical composition on a DM basis of the total mixed ration.

Ingredient	Composition (%)
Yellow maize	28.0
Eragrostis curvula hay	22.2
Alfalfa hay	20.0
Soybean meal	17.0
Molasses	6.0
Wheat	5.0
Urea	0.8
Vitamin premix	0.5
Parameter	*Chemical composition* (%)
Dry matter	89.7
CP	17.2
Ash	6.5
Starch	6.5
NDF	3.4
ADF	24.2
Lignin	2.5
ME	0.9

All experimental animals received the same total mixed ration (TMR) during the experimental period. The TMR used was formulated to support an average daily gain (ADG) of approximately 250 g/head/day following the Agricultural Research Council's [17] recommendations. The following four dietary additives were formed:

- TMR only: (control treatment);
- TMR plus *A. indica* leaf extract at a dosage of 50 mg per kg feed: (neem treatment);
- TMR plus *M. oleifera* leaf extract at a dosage of 50 mg per kg feed: (moringa treatment);
- TMR plus monensin sodium at a dosage of 50 mg per kg of feed: (monensin; positive control treatment).

The lambs from all treatment groups were fed ad libitum, and clean water was available ad libitum. Feed intake was calculated by subtracting the amount of refused feed from the feed offered the day before. A random sample was collected from the amount of feed offered each day as a retention sample for feed analysis later. The *A. indica* and *M. oleifera* extracts were reconstituted with distilled water (1 g extract to 1 L distilled water) and administered. The solutions of plant extract were administered at a dosage of 50 mL per kg of feed DM as recommended by Akanmu and Hassen [14]. The average DMI of the lambs was in the range of 1655 to 1866 g/head/day. The required dosages were drenched to lambs in the morning and afternoon before feeding using a 20 mL metal drencher (NJ Philips, Somersby, Australia).

The initial body weights of the animals were recorded for three consecutive days before the start of the experiment and thereafter at seven-day intervals before the morning feeding until the end of the experimental period. The final weights of the animals were also recorded for three consecutive days before the morning feeding. The lambs were reared over a 23 week trial period to a marketable weight of 60–65 kg (9–10 months old). The data on feed intake, nutrient digestibility, average daily gain and methane measurement were document by Du Preez [18]. The lambs were slaughtered when they reached the required final weight of 48–52 kg, and the meat samples collected were used for the current study.

2.3. Slaughter and Sampling Procedure

The lambs were slaughtered according to the standard procedure at the Renbro Abattoir, Hammanskraal, South Africa. Carcasses were immediately weighed to obtain the hot carcass weight and classified using the South African Carcass Classification System for beef, sheep and goat carcasses. This carcass classification system classifies carcasses based on their physical and compositional attributes, which include age (age categories: A, AB, B and C), carcass fatness (carcass fat codes: 1 to 6), carcass conformation (carcass conformation codes: 1 to 5) and damage (1 to 3) [19]. The carcasses were then chilled at 4 °C for 24 h.

After 24 h in the chilling room, the carcasses were reweighed to obtain the cold carcass weight, and they were transferred to the laboratory for dissection under refrigerated conditions. Carcass composition was determined using the method described by Casey et al. [20]. Briefly, a three-rib sample was cut from the 8th, 9th and 10th lumbar vertebrae on the left side of each carcass, the ventral extremity of the sample being on a line drawn from the pubic symphysis to the middle of the first rib [20]. The three-rib cut sample was dissected into meat, fat and bone to obtain an estimate of the total carcass composition. The meat, fat and bone carcass components were vacuum packed and stored in the freezer at −20 °C until further analysis. Subcutaneous fat (SCF) and intramuscular fat (IMF) samples of approximately 5 g each were dissected from the three-rib-cut *Longissimus* muscle and stored in polythene bags at −20 °C for fatty acid analysis.

2.4. Analytical Procedures

The dry matter content and ether extract of the *Longissimus* muscle samples were determined with the method used by the Association of Official Analytical Chemists [15]. The method involved boiling about 1 g of freeze-dried meat samples in petroleum ether for two hours and then oven drying until all the petroleum ether had evaporated. Thereafter, the samples were weighed and expressed as a percentage of the whole sample. Fat pigments were extracted according to the method of Kirton et al. [21]. The absorbance of each fraction was measured in a spectrophotometer (Specord 200®) at a 423 nm wavelength, and the lutein concentrations were calculated using Beer's law equation [22].

2.5. Fatty Acid Analysis

The lipid extraction procedure and determination of fatty acid methyl esters were described by Webb and Casey [23]. Briefly, lipids were extracted in duplicate using a modification [24] of the chloroform:methanol (2:1, v/v) method [25]. Butylated hydroxytoluene (2.6 Di-tert-Butyl-p-Cresol) was included as an antioxidant. Methyl esters of the fatty acid component of the neutral triglycerides were prepared according to the NaOH/methanol method [15]. These esters were separated on a polar phase SP2330 column (2 m × 3 mm, packed with Silar 1OC coated on a Gas Chrom Q) fitted to a Shimadzu Tracera gas chromatograph with a barrier ionisation discharge detector. Profiles of the cis–trans fatty acids from the subcutaneous adipose tissues were obtained from fat samples that were treated with n-hexane at 35 °C for 24 h, after which the fatty acids were esterified according to the method of Van Wijngaarden [26]. The cis–trans fatty acids isomers were then separated on an SP2560 fused silica capillary column (100 m × 0.2 mm) fitted to a Varian 3700 gas chromatograph. Standards for the fatty acids were obtained from Nu-Chek-Prep., Inc. (Elysim, MI, USA). Fatty acids were expressed in both normalised (i.e., molar proportion) and gravimetric (i.e., milligrams per gram of fresh tissue) formats [27,28].

2.6. Statistical Methods

Data were analysed in a randomised complete block design. The variables of the carcass' characteristics, lutein concentrations and meat fatty acids were first tested for normality and homoscedasticity with the Shapiro–Wilk and Levene's tests, respectively. Statistical analysis was performed using the general linear model (GLM) ANOVA procedure in SPSS version 27, and the model included the treatment effect. Differences were considered significant at $p < 0.05$ and a tendency for significance at $0.05 < p < 0.10$. The post hoc analyses were conducted with the Bonferroni comparison procedure in SPSS version 27. Factor component scores (z-scores) were calculated for the three-rib cut fat content to calculate the carcass fat content (CFC), after which the z-scores were transformed to standard scores (t-scores). Principal component factor analysis (PCA) was used to compute a succinct factor index for the carcass fat content (CFC t-score) and to describe the main effect of the feed additive treatments on the parameters measured. The measurements of the PCA plots were interpreted according to the correlations between each parameter. On the PCA plot, measurements close together are positively correlated; measurements separated by 180° are negatively correlated and measurements separated by 90° are independent [29].

3. Results

3.1. Carcass Characteristics

The results of the carcass characteristics of the lambs fed diets supplemented with neem (*A. indica*) and moringa (*M. oleifera*) leaf extracts are presented in Table 2.

Table 2. The effects of dietary inclusion of neem (*Azadirachta indica*) and moringa (*Moringa oleifera*) leaf extract on the carcass characteristics (LS means ± SE) of South African Mutton Merino lambs.

Treatment	Control	Neem	Moringa	Monensin	*p*-Value
Initial weight (kg)	38.6 ± 0.95	38.3 ± 1.29	38.1 ± 1.17	37.4 ± 1.43	0.95
CCW (kg)	29.2 ± 0.55	30.2 ± 0.65	30.8 ± 1.31	28.7 ± 0.92	0.61
Meat (%)	51.8 ± 1.11	51.6 ± 0.81	55.5 ± 0.86	54.4 ± 0.70	0.06
Fat (%)	33.2 ± 1.22	34.3 ± 0.77	30.1 ± 1.23	30.0 ± 0.85	0.07
Bone (%)	15.0 ± 0.40	14.1 ± 0.72	14.4 ± 0.61	15.6 ± 0.48	0.51
LM dry matter (%)	32.7 ± 0.63	34.5 ± 0.90	34.5 ± 1.70	35.8 ± 2.44	0.11
IMF (%)	12.0 ± 0.89	15.2 ± 0.71	12.2 ± 1.33	12.6 ± 0.97	0.28
CFC t-scores	50.3± 2.43	58.9 ± 2.15	45.4 ± 4.11	46.2 ±2.55	0.05

CCW: cold carcass weight; LM: longissimus muscle; IMF: intramuscular fat; CFC: carcass fat content.

There were no significant differences in cold carcass weight among the dietary treatment groups. However, there was a tendency for a treatment effect on meat percentage ($p = 0.06$), fat percentage ($p = 0.07$) and CFC *T*-scores ($p = 0.05$) but the differences were not significant. Dietary treatment had no effect ($p = 0.11$) on dry matter and IMF content of the *Longissimus* muscle.

3.2. Pigmentation of the Subcutaneous Fat

The lutein pigment concentration in the subcutaneous fat of the lambs supplemented with neem, moringa and monensin compared to the control are presented in Table 3.

Table 3. The effect of dietary inclusion of neem (*Azadirachta indica*) and moringa (*Moringa oleifera*) on the lutein concentrations in the subcutaneous fat of SA Mutton Merino lambs.

	Control	Neem	Moringa	Monensin	*p*-Value
Lutein (mg/100 g)	1.16 ^{a} ± 0.14	0.59 ^{b} ± 0.13	0.86 ^{ab} ± 0.14	0.57 ^{b} ± 0.14	0.03

^{ab} Means with different superscripts were significantly different ($p < 0.05$).

The content of the lutein pigment was affected by the dietary treatment ($p = 0.03$). The content of the lutein pigment was significantly higher ($p < 0.05$) in lambs fed with the control diet compared to those supplemented with neem or monensin ($p < 0.05$), while those fed with moringa had an intermediate ($p > 0.05$) lutein content. The practical implication is that the subcutaneous fat from the control (i.e., nonsupplemented) lambs was slightly more yellow compared to those supplemented with neem or monensin.

3.3. Fatty Acid Composition

The molar proportions of fatty acids in the SCF of lambs are presented in Table 4. The saturated fatty acids (SFAs) comprised 49.7–54.3% of the total fatty acids in the subcutaneous fat of the lambs. The main SFAs were palmitic acid (C16:0; 26–27.6%) and stearic acid

(C18:0; 15.7–19.2%). There was a tendency toward a treatment effect on the proportion of stearic acid ($p = 0.05$). Although supplementation with the plant extracts did not affect the proportion of margaric acid (C17:0), docosanoic acid (C22:0) and tricosanoic acid (C23:0), monensin supplementation significantly decreased the C17:0 content and increased the proportion of both C22:0 and C23:0 compared to the control treatment ($p < 0.05$).

Table 4. The effects of dietary inclusion of *Azadirachta indica* and *Moringa oleifera* leaf extract on the fatty acid composition of the subcutaneous fat (w/w%; LS mean ± SE) of South African Mutton Merino lambs.

Fatty Acids	Control	Neem	Moringa	Monensin	*p*-Value
C14:0	3.09 ± 0.28	3.20 ± 0.22	2.76 ± 0.15	3.45 ± 0.24	0.43
C16:0	27.0 ± 0.66	27.6 ± 0.64	26.0 ± 0.82	26.6 ± 0.45	0.59
C17:0	5.52 ^{a} ± 0.44	3.74 ^{ab} ± 0.33	4.52 ^{ab} ± 0.15	3.59 ^{b} ± 0.44	0.03
C18:0	16.7 ± 1.31	17.6 ± 0.65	15.7 ± 0.64	19.2 ± 1.39	0.05
C20:0	0.11 ± 0.01	0.11 ± 0.01	0.11 ± 0.01	0.13 ± 0.01	0.37
C21:0	0.51 ± 0.03	0.41 ± 0.03	0.47 ± 0.05	0.53 ± 0.08	0.65
C22:0	0.02 ^{a} ± 0.00	0.03 ^{ab} ± 0.00	0.03 ^{ab} ± 0.00	0.04 ^{b} ± 0.00	0.04
C23:0	0.05 ^{a} ± 0.00	0.07 ^{ab} ± 0.01	0.06 ^{ab} ± 0.01	0.09 ^{b} ± 0.01	0.04
SFA	50.1 ^{a} ±1.01	52.8 ^{b} ± 0.55	49.7 ^{a} ± 0.69	54.3 ^{b} ± 1.16	0.03
C14:1	0.25 ± 0.03	0.17 ± 0.02	0.18 ± 0.01	0.15 ± 0.02	0.13
C15:1	0.03 ± 0.01	0.02 ± 0.01	0.03 ± 0.01	0.03 ± 0.01	0.85
C16:1	1.32 ± 0.13	1.07 ± 0.09	1.08 ± 0.08	0.96 ± 0.06	0.23
C18:1n9c	44.4 ^{ab} ± 1.05	42.4 ^{ab} ± 0.76	45.0 ^{a} ± 0.57	40.5 ^{b} ± 0.80	0.02
C18:1n9t	0.79 ± 0.15	0.59 ± 0.07	0.79 ± 0.17	0.95 ± 0.20	0.65
C20:1	0.15 ^{a} ± 0.01	0.08 ^{b} ± 0.00	0.14 ^{ab} ± 0.02	0.09 ^{ab} ± 0.01	0.02
MUFAs	46.9 ^{ab} ± 1.02	44.1 ^{ab} ± 0.69	47.3 ^{a} ± 0.66	42.6 ^{b} ± 0.87	0.02
C20:2	0.08 ± 0.00	0.07 ± 0.00	0.08 ± 0.00	0.08 ± 0.01	0.76
C18:2n6c	2.05 ^{a} ± 0.04	2.07 ^{a} ± 0.15	2.16 ^{ab} ± 0.11	2.71 ^{b} ± 0.11	0.01
C18:3n3	0.36 ^{a} ± 0.01	0.38 ^{a} ± 0.02	0.39 ^{a} ± 0.02	0.51 ^{b} ± 0.03	<0.01
C18:3n6	0.03 ± 0.00	0.03 ± 0.00	0.03 ± 0.00	0.04 ± 0.00	0.10
C20:3n6	0.03 ± 0.00	0.03 ± 0.00	0.03 ± 0.00	0.03 ± 0.00	0.38
PUFAs	2.53 ^{a} ± 0.06	2.56 ^{a} ± 0.18	2.66 ^{ab} ± 0.14	3.36 ^{b} ± 0.13	<0.01
UFA/SFA	0.99 ^{a} ± 0.04	0.89 ^{ab} ± 0.02	1.01 ^{a} ± 0.03	0.85 ^{b} ± 0.03	0.03
PUFA n-6/n-3	5.86 ± 0.13	5.47 ± 0.23	5.61 ± 0.15	5.56 ± 0.31	0.78
PUFA/SFA	0.05 ^{a} ± 0.00	0.05 ^{a} ± 0.00	0.05 ^{a} ± 0.00	0.06 ^{b} ± 0.00	0.05

^{ab} Means with different superscripts were significantly different ($p < 0.05$). SFAs: saturated fatty acids; MUFAs: monounsaturated fatty acids; PUFAs: polyunsaturated fatty acids.

The molar proportion of the monounsaturated fatty acids (MUFAs) accounted for approximately 45% of the total fatty acids. The major MUFA was oleic acid (C18:1n9c) which comprised approximately 43% of the total fatty acids and approximately 96% of MUFAs. The proportion of oleic acid (C18:1n9c) was affected ($p = 0.02$) by dietary treatment, since a significant difference was observed in the proportion of oleic acid (C18:1n9c) between the moringa and monensin treatment groups (45.0% vs. 40.5%). Similarly, total MUFAs were higher in the moringa treatment group compared to the monensin treatment group (47.3% vs. 42.6%).

Total polyunsaturated fatty acids (PUFAs) accounted for approximately 2.8% of total fatty acids. Linoleic acid (C18:2n6c) was the main PUFA in the SCF of lambs in all groups. Dietary treatment affected total PUFAs ($p < 0.01$). Lambs supplemented with monensin had higher PUFAs in comparison to the control (2.53%; $p < 0.05$) and neem (2.56%; $p < 0.05$) treatment groups. This dietary treatment effect was the result of higher ($p < 0.05$) proportions of linoleic acid (C18:2n6c) and α-linoleic acid (C18:3n3) in the SCF of lambs supplemented with monensin (2.71%) compared to the control and neem treatment groups.

Dietary treatment affected the UFA:SFA ratio (p = 0.03). Lambs in the monensin treatment group had lower ($p < 0.05$) UFA:SFA ratios compared to the control group and moringa treatment group. The n-6/n-3 (p = 0.78) and PUFA: SFA (p = 0.05) ratios were not affected by the dietary additives.

3.4. Overview of the Feed Additive Treatment's Main Effects on Physiological Parameters

The results of the PCA of the parameters considered in this study are presented in Figure 1. PC 1 (i.e., fatty acid composition component) is presented on the x-axis, and PC 2 (i.e., fat content component) is presented on the y-axis. The carcass fat content t-score, fat%, IMF% and days-on-trial showed a strong positive correlation with PC 2 (i.e., fat content component). Considering the direction of the PCA plot projections, it is evident that dietary supplementation with neem considerably increased the carcass fat content compared to the control and all of the other feed additive treatment groups.

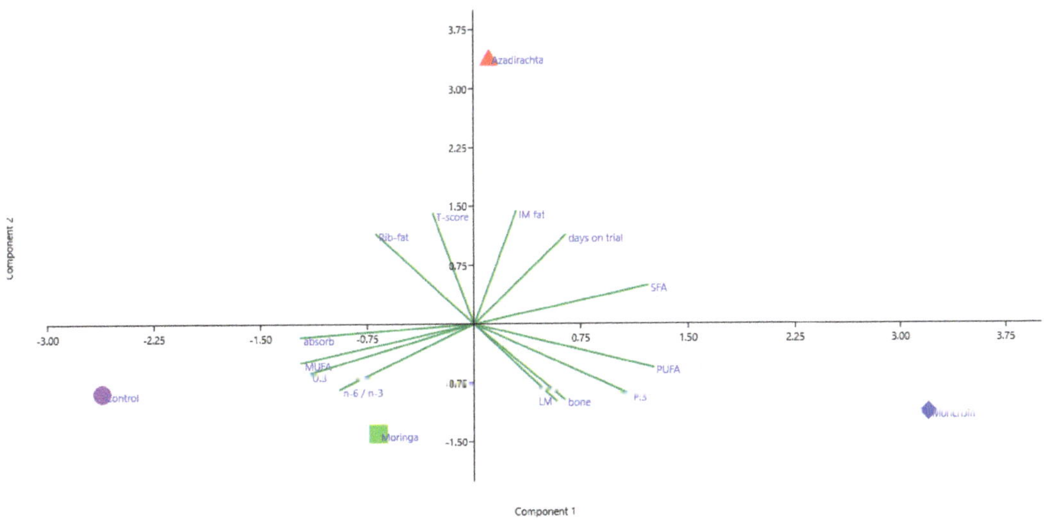

Figure 1. Projections of carcass and meat quality measurements on a plane defined by PC 1 (i.e., fatty acid composition component) and PC 2 (i.e., fat content component). Fat%: Rib-fat; carcass fat content t-scores: T-score); intramuscular fat%: IM fat; saturated fatty acids: SFA; PUFA:SFA—P:S; bone percentage: bone; meat%: LM; UFA:SFA—U:S; absorbance: absorb. Neem treatment: Azadirachta; monensin treatment: monensin; moringa treatment: moringa; negative control treatment: control.

On the other hand, the proportions of PUFAs, SFAs and PUFA:SFA showed the largest positive correlations with PC 1 (i.e., fatty acid composition component), emphasising the improvements gained in terms of favourable fatty acids (i.e., MUFAs and UFA:SFA) by the supplementation of lambs' diet with moringa and the control treatment compared to both the monensin and neem feed treatment groups. However, lambs supplemented with monensin showed an increase in PUFAs, PUFA:SFA, meat% and bone% more than the other treatment groups.

4. Discussion

In the present study, lambs were fed to weights exceeding normal market weight (e.g., 60–65 kg) over a 23 week trial period to study the effects of plant supplements on carcass fat content and composition. The cold carcass weights were similar to those recorded in South African Mutton Merino (SAMM) lambs slaughtered at the same age/weight [30,31]. The meat percentage was lower, while the fat percentage was higher than previously reported for other South African lamb breeds in the A-age class [32] due to the differences in the slaughter age/weight. Future studies should consider slaughtering lambs when they reach a normal market weight, approximately 45 to 50 kg, for an ideal carcass quality that conforms to consumer preferences [30,33].

In terms of carcass composition, supplementing high-fibre diets with *A. indica* and *M. oleifera* leaf extract as antimethanogenic agents did not affect the meat and fat contents of the lamb carcasses. Previous studies have shown that the inclusion of *M. oleifera* in diets, shifts rumen fermentation kinetics from acetate to propionate, a major precursor of glucose synthesis in the liver that may be subsequently used for protein biosynthesis [34]. This effect is attributed to secondary bioactive compounds in *M. oleifera* plant extract, which have been reported to inhibit Gram-positive bacteria and favour propionate-producing bacteria [35]. The present results on carcass quality suggest that both *A. indica* and *M. oleifera* could be used in lamb diets as antimethanogenic additives, with the advantage of being natural additives that have no negative effects on carcass fatness.

The degree of marbling is an important attribute of carcasses, and it is used as a visual cue by consumers to judge the quality of meat [36]. In the present study, the IMF values were within the range previously reported for lambs [37]. The IMF percentages of lambs were not significantly different across the four dietary treatment groups. The inclusion of *A. indica* and *M. oleifera* leaf extracts in lamb diets neither improved nor compromised the visual appearance of the meat.

Yellow carcass fat is negatively evaluated by consumers in many countries [38]. Fat colour changes from a creamy white to a bright yellow–orange with the accumulation of carotenoids [22]. It is widely accepted that lutein is the main carotenoid in sheep adipose tissue [22,39,40]. Studies that quantify carotenoids in sheep fat are very scarce, because the concentrations are very low compared to carotenoids in cattle. The lutein concentrations found in the present study (0.57–1.16 mg/100 g feed) compare well to values previously reported in the literature for South African lamb breeds [33].

Previous studies have reported that plant secondary compounds have a protective effect on carotenoids resulting in higher depositions [41]. This effect was not observed in this study in both of our experimental treatment groups, which showed numerically lower lutein values compared to the control treatment. We can only speculate that low lutein concentrations in the neem and monensin treatment groups observed in this study were possibly related to their IMF content. Research has shown that as carotenoids accumulate in adipocytes, the increase in IMF may dilute the carotenoids and, consequently, reduce the yellowness of the subcutaneous fat [42]. More studies of the effect of medicinal plant extracts on the deposition of carotenoids in the adipose tissue of lambs/sheep should be considered.

Fat and long-chain fatty acids contribute to important aspects of meat quality and are key to the nutritional and sensory values of the meat [43]. In the present study, SFAs constituted approximately half of the total fatty acids in SCF of lambs, typical of SAMM lambs kept on high forage diets [23]. This is related to the fact that forages stimulate ruminal activity and biohydrogenation of fatty acids thus increasing the proportion of SFAs [44]. The main SFAs were palmitic acid (C16:0) and stearic acid (C18:0). This prevalence is in line with previously reported values for SAMM lambs kept on high forage-based diets [31].

Although SFAs are generally considered unhealthy, some have positive benefits on human health. It is only myristic acid (C14:0) and palmitic acid (C16:0) that are associated with an increased risk of obesity, hypercholesterolemia, some cancers and a decrease in LDL cholesterol [43,45]. Overall, our experimental diets did not alter the SFA proportion. The only SFAs affected by the dietary plant supplements were margaric acid (C17:0), docosanoic acid (C22.0) and tricosanoic acid (C23:0), and the difference was between the monensin and control treatment, while the moringa and neem treatments showed intermediate values. Our results indicate that the inclusion of *M. oleifera* and *A. indica* leaf extracts in lamb diets does not affect the SFA content of the meat and, by implication, does not cause increased risk to human health as previously suggested [46].

Oleic acid (C18:1n9c) was the main MUFA and conjugated linoleic acid (C18:2n6c) was the most prominent PUFA, as previously reported for lamb meat [45]. PUFAs (n-3 and n-6) are generally regarded as beneficial for human health [47]. However, high proportions of PUFAs can have negative effects on quality aspects such as fat firmness, shelf life and meat flavour [33]. In the present study higher proportions of oleic acid (C18:1n9c) were deposited in the SCF of lambs supplemented with moringa plant extract compared to monensin, which could be beneficial for human health. Higher proportions of linoleic acid (C18.2n6c) (n-6) were deposited in the SCF of lambs fed monensin compared to the neem and control treatment groups. Higher proportions and alpha-linoleic acid (C18:3n3) (n-3) were deposited in the SCF of lambs fed monensin-supplemented diets as opposed to other dietary treatment groups. However, the differences among dietary treatments in the proportions of PUFAs deposited in the SCF were minor (less than 1%) and will presumably not have any impact on human health, organoleptic properties and the technological quality of the meat.

The UFA:SFA ratio is commonly used to assess the nutritional value of fats, while the PUFA n-6-to-n-3 ratio indicates the risk of coronary heart disease or cancer in humans. In the present study, both the UFA:SFA and n-6/n-3 ratios were above the minimum recommended values of 0.4 and 4, respectively [43,47]. The differences across dietary treatments were not significant and, hence, were presumably of minor importance.

5. Conclusions

Despite small differences, the inclusion of *M. oleifera*, as a feed additive, resulted in higher oleic acid, MUFAs and UFA:SFA ratio compared to the monensin treatment, which could be considered beneficial for human health. *A. indica* had no or minimal effects on the carcass' characteristics and the meat fatty acid composition. Overall, the antimethanogenic feed additives investigated in this study had no negative effects on the carcass fat and fatty acid composition of the lambs. Therefore, *A. indica* and *M. oleifera* feed additives can be used as safe and inexpensive antimethanogenic agents, without compromising the resultant meat quality.

Author Contributions: Conceptualisation, A.H. and E.C.W.; methodology, A.H. and E.C.W.; software, E.C.W.; validation, E.C.W., A.H. and M.O.O.; formal analysis, E.C.W.; investigation, M.O.O.; resources, A.H. and E.C.W.; data curation, E.C.W., writing—original draft preparation, M.O.O. and P.P.; writing—review and editing, E.C.W., A.H. and P.P.; visualisation, E.C.W. and A.H.; supervision, A.H. and E.C.W.; project administration, A.H.; funding acquisition, A.H. All authors have read and agreed to the published version of the manuscript.

Funding: This research received National Research Foundation (NRF) funding under grant numbers: 98903 and 118518.

Institutional Review Board Statement: The animal study protocol was approved by the Animal Ethics Committee of the University of Pretoria (protocol number: EC004-18; approved on 30 January 2018).

Informed Consent Statement: Not applicable.

Data Availability Statement: Data supporting the reported results are available upon request.

Acknowledgments: We would like to thank the Queen Elizabeth Commonwealth Scholarships for financial assistance provided to the first author and the National Research Foundation (NRF) for funding the running cost of the research under grant number: 98903 and 118518.

Conflicts of Interest: The authors declare no conflict of interest.

References

1. Gustafson, R.H. Use of antibiotics in livestock and human health concerns. *J. Dairy Sci.* **1991**, *74*, 1428–1432. [CrossRef]
2. Wegener, H.C. Antibiotics in animal feed and their role in resistance development. *Curr. Opin. Microbiol.* **2003**, *6*, 439–445. [CrossRef] [PubMed]
3. Phillips, I. Withdrawal of growth-promoting antibiotics in Europe and its effects in relation to human health. *Int. J. Antimicrob. Agents* **2007**, *30*, 101–107. [CrossRef] [PubMed]
4. Scott, A.M.; Beller, E.; Glasziou, P.; Clark, J.; Ranakusuma, R.W.; Byambasuren, O.; Bakhit, M.; Page, S.W.; Trott, D.; Del Mar, C. Is antimicrobial administration to food animals a direct threat to human health? A rapid systematic review. *Int. J. Antimicrob. Agents* **2018**, *52*, 316–323. [CrossRef]
5. Olagaray, K.E.; Bradford, B.J. Plant flavonoids to improve productivity of ruminants—A review. *Anim. Feed Sci. Technol.* **2019**, *251*, 21–36. [CrossRef]
6. Akanmu, A.M.; Hassen, A.; Adejoro, F.A. Gas production, digestibility and efficacy of stored or fresh plant extracts to reduce methane production on different substrates. *Animals* **2020**, *10*, 146. [CrossRef]
7. Vasta, V.A.; Daghio, M.A.; Cappucci, A.L.; Buccioni, A.R.; Serra, A.; Viti, C.A.; Mele, M.A. Invited review: Plant polyphenols and rumen microbiota responsible for fatty acid biohydrogenation, fiber digestion, and methane emission: Experimental evidence and methodological approaches. *J. Dairy Sci.* **2019**, *102*, 3781–3804. [CrossRef]
8. Kholif, A.E.; Olafadehan, O.A. Dietary strategies to enrich milk with healthy fatty acids—A review. *Ann. Anim. Sci.* **2021**, *22*, 523–536. [CrossRef]
9. Jack, A.A.; Oghenesuvwe, O.; Adewumi, M.K.; Omojola, A.B.; Adegbeye, M.J.; Faniyi, T.O.; Salem, A.Z.M.; Elghandour, M.M.M.Y.; Cuevas-Barragán, C.E.; Barbabosa-Pliego, A.; et al. Conversion of Neem fruit biomass for rumen manipulation, meat fatty acid profile improvement of rams. *Biomass Convers. Biorefin.* **2022**. [CrossRef]
10. Cohen-Zinder, M.; Orlov, A.; Trofimyuk, O.; Agmon, R.; Kabiya, R.; Shor-Shimoni, E.; Wagner, E.K.; Hussey, K.; Leibovich, H.; Miron, J. Dietary supplementation of Moringa oleifera silage increases meat tenderness of Assaf lambs. *Small Rumin. Res.* **2017**, *151*, 110–116. [CrossRef]
11. Babiker, E.E.; Juhaimi, F.A.; Ghafoor, K.; Abdoun, K.A. Comparative study on feeding value of Moringa leaves as a partial replacement for alfalfa hay in ewes and goats. *Livest. Sci.* **2017**, *195*, 21–26. [CrossRef]
12. Jena, A.K.; Karan, M.; Vasisht, K. Plant parts substitution based approach as a viable conservation strategy for medicinal plants: A case study of Premna latifolia Roxb. *J. Ayurveda Integr. Med.* **2017**, *8*, 68–72. [CrossRef] [PubMed]
13. McGaw, L.J.; Eloff, J.N. Ethnoveterinary use of southern African plants and scientific evaluation of their medicinal properties. *J. Ethnopharmacol.* **2008**, *119*, 559–574. [CrossRef] [PubMed]
14. Akanmu, A.M.; Hassen, A. The use of certain medicinal plant extracts reduced in vitro methane production while improving in vitro organic matter digestibility. *Anim. Prod. Sci.* **2017**, *58*, 900–908. [CrossRef]
15. AOAC. *Official Methods of Analysis of AOAC International*; AOAC: Gaithersburg, MD, USA, 2000.
16. Van Soest, P.J.; Robertson, J.B.; Lewis, B.A. Methods for dietary fiber, neutral detergent fiber, and non-starch polysaccharides in relation to animal nutrition. *J. Dairy Sci.* **1991**, *74*, 3583–3597. [CrossRef]
17. Agricultural Research Council. *The Nutrient Requirements of Ruminants Livestock*; Commonwealth Agricultural Bureaux: London, UK, 1980.
18. Du Preez, D.A. The Effect of *Azadirachta indica* and *Moringa oleifera* on Nutrient Digestibility, Growth Performance and Methanogenesis in SA Mutton Merino sheep. Master's Dissertation, University of Pretoria, Pretoria, South Africa, 24 April 2020.
19. Webb, E.C. Description of carcass classification goals and the current situation in South Africa. *S. Afr. J. Anim. Sci.* **2015**, *45*, 229–233. [CrossRef]
20. Casey, N.H.; van Niekerk, W.A.; Spreeth, E.B. Fatty acid composition of subcutaneous fat of sheep grazed on eight different pastures. *Meat Sci.* **1988**, *23*, 55–63. [CrossRef]
21. Kirton, A.H.; Crane, B.; Paterson, D.J.; Clare, N.T. Yellow fat in lambs caused by carotenoid pigmentation. *N. Z. J. Agric. Res.* **1975**, *18*, 267–272. [CrossRef]
22. Kruggel, W.G.; Field, R.A.; Miller, G.J.; Horton, K.M.; Busboom, J.R. Influence of sex and diet on lutein in lamb fat. *J. Anim. Sci.* **1982**, *54*, 970–975. [CrossRef]
23. Webb, E.C.; Casey, N.H. Genetic differences in fatty acid composition of subcutaneous adipose tissue in Dorper and SA Mutton Merino wethers at different live weights. *Small Rumin. Res.* **1995**, *18*, 81–88. [CrossRef]
24. Ways, P.; Hanahan, D.J. Characterization and quantification of red cell lipids in normal man. *J. Lipid Res.* **1964**, *5*, 318–328. [CrossRef]
25. Folch, J.; Lees, M.; Stanley, G.S. A simple method for the isolation and purification of total lipides from animal tissues. *J. Biol. Chem.* **1957**, *226*, 497–509. [CrossRef]
26. Van Wijngaarden, D. Modified rapid preparation of fatty acid esters from lipids for gas chromatographic analysis. *Anal. Chem.* **1967**, *39*, 848–849. [CrossRef]
27. Slover, H.T.; Lanza, E. Quantitative analysis of food fatty acids by capillary gas chromatography. *J. Am. Oil Chem. Soc.* **1979**, *56*, 933–943. [CrossRef]
28. Huerta-Leidenz, N.O.; Cross, H.R.; Savell, J.W.; Lunt, D.K.; Baker, J.F.; Pelton, L.S.; Smith, S.B. Comparison of the fatty acid composition of subcutaneous adipose tissue from mature Brahman and Hereford cows. *J. Anim. Sci.* **1993**, *71*, 625–630. [CrossRef]

29. Cañeque, V.; Pérez, C.; Velasco, S.; Dıaz, M.T.; Lauzurica, S.; Álvarez, I.; de Huidobro, F.R.; Onega, E.; de la Fuente, J. Carcass and meat quality of light lambs using principal component analysis. *Meat Sci.* **2004**, *67*, 595–605. [CrossRef]
30. Brand, T.S.; van der Westhuizen, E.J.; van der Merwe, D.A.; Hoffman, L.C. Effect of days in feedlot on growth performance and carcass characteristics of Merino, South African Mutton Merino and Dorper lambs. *S. Afr. J. Anim. Sci.* **2017**, *47*, 26–33. [CrossRef]
31. Webb, E.C.; Casey, N.H.; van Niekerk, W.A. Fatty acids in the subcutaneous adipose tissue of intensively fed SA Mutton Merino and Dorper wethers. *Meat Sci.* **1994**, *38*, 123–131. [CrossRef]
32. Tshabalala, P.A.; Strydom, P.E.; Webb, E.C.; de Kock, H.L. Meat quality of designated South African indigenous goat and sheep breeds. *Meat Sci.* **2003**, *65*, 563–570. [CrossRef]
33. Webb, E.C.; Casey, N.H.; Bosman, M.J.C. Dietary influences on lutein pigments and carcass fat quality in wethers of different maturity types. *S. Afr. J. Anim. Sci.* **1999**, *29*, 83–91. [CrossRef]
34. Kholif, A.E.; Gouda, G.A.; Olafadehan, O.A.; Abdo, M.M. Effects of replacement of Moringa oleifera for berseem clover in the diets of Nubian goats on feed utilisation, and milk yield, composition and fatty acid profile. *Animal* **2018**, *12*, 964–972. [CrossRef] [PubMed]
35. Soltan, Y.A.; Hashem, N.M.; Morsy, A.S.; El-Azrak, K.M.; El-Din, A.N.; Sallam, S.M. Comparative effects of *Moringa oleifera* root bark and Monensin supplementations on ruminal fermentation, nutrient digestibility and growth performance of growing lambs. *Anim. Feed Sci. Technol.* **2018**, *235*, 189–201. [CrossRef]
36. Realini, C.E.; Pavan, E.; Johnson, P.L.; Font-i-Furnols, M.; Jacob, N.; Agnew, M.; Craigie, C.R.; Moon, C.D. Consumer liking of M. longissimus lumborum from New Zealand pasture-finished lamb is influenced by intramuscular fat. *Meat Sci.* **2021**, *173*, 108380. [CrossRef]
37. Alizadeh, A.; Shahneh, A.Z.; Yousefi, A.R.; Omran, M.H.; Campbell, A.W. Determining the effect of the fat-tail and carcass weight on meat fatty acid composition of Iranian lambs. *Small Rumin. Res.* **2013**, *115*, 34–39. [CrossRef]
38. Troy, D.J.; Kerry, J.P. Consumer perception and the role of science in the meat industry. *Meat Sci.* **2010**, *86*, 214–226. [CrossRef] [PubMed]
39. Yang, A.; Larsen, T.W.; Tume, R.K. Carotenoid and retinol concentrations in serum, adipose tissue and liver and carotenoid transport in sheep, goats and cattle. *Aust. J. Agric. Res.* **1992**, *43*, 1809–1817. [CrossRef]
40. Dian, P.H.M.; Andueza, D.; Jestin, M.; Prado, I.N.; Prache, S. Comparison of visible and near infrared reflectance spectroscopy to discriminate between pasture-fed and concentrate-fed lamb carcasses. *Meat Sci.* **2008**, *80*, 1157–1164. [CrossRef]
41. Rufino-Moya, P.J.; Joy, M.; Lobón, S.; Bertolín, J.R.; Blanco, M. Carotenoids and liposoluble vitamins in the plasma and tissues of light lambs given different maternal feedings and fattening concentrates. *Animals* **2020**, *10*, 1813. [CrossRef]
42. Torrecilhas, J.A.; Ornaghi, M.G.; Passetti, R.A.C.; Mottin, C.; Guerrero, A.; Ramos, T.R. Meat quality of young bulls finished in a feedlot and supplemented with clove or cinnamon essential oils. *Meat Sci.* **2021**, *174*, 108412. [CrossRef]
43. Webb, E.C.; O'Neill, H.A. The animal fat paradox and meat quality. *Meat Sci.* **2008**, *80*, 28–36. [CrossRef]
44. Choi, N.J.; Imm, J.Y.; Oh, S.; Kim, B.C.; Hwang, H.J.; Kim, Y.J. Effect of pH and oxygen on conjugated linoleic acid (CLA) production by mixed rumen bacteria from cows fed high concentrate and high forage diets. *Anim. Feed Sci. Technol.* **2005**, *123*, 643–653. [CrossRef]
45. Hajji, H.; Joy, M.; Ripoll, G.; Smeti, S.; Mekki, I.; Gahete, F.M.; Mahouachi, M.; Atti, N. Meat physicochemical properties, fatty acid profile, lipid oxidation and sensory characteristics from three North African lamb breeds, as influenced by concentrate. *J. Food Compost. Anal.* **2016**, *48*, 102–110. [CrossRef]
46. Falowo, A.B.; Mukumbo, F.E.; Idamokoro, E.M.; Lorenzo, J.M.; Afolayan, A.J.; Muchenje, V. Multi-functional application of Moringa oleifera Lam. in nutrition and animal food products: A review. *Food Res. Int.* **2018**, *106*, 317–334. [CrossRef] [PubMed]
47. Scollan, N.; Hocquette, J.F.; Nuernberg, K.; Dannenberger, D.; Richardson, I.; Moloney, A. Innovations in beef production systems that enhance the nutritional and health value of beef lipids and their relationship with meat quality. *Meat Sci.* **2006**, *74*, 17–33. [CrossRef] [PubMed]

animals

Communication

Carcass Traits of Growing Meat Goats Fed Different Levels of Hempseed Meal

Reshma Gurung [1], Khim B. Ale [1], Frank W. Abrahamsen [1], Katie Moyer [2], Jason T. Sawyer [3] and Nar K. Gurung [1,*]

[1] Department of Agricultural and Environmental Sciences, Tuskegee University, Tuskegee, AL 36088, USA
[2] Kentucky for Hemp, Crofton, KY 42217, USA
[3] Department of Animal Sciences, Auburn University, Auburn, AL 36849, USA
* Correspondence: ngurung@tuskegee.edu; Tel.: +1-334-727-8457

Simple Summary: Industrial hemp is currently being investigated as a potential new crop in the U.S. after the passage and approval of the 2014 and 2018 Farm Bills. Hemp plants grow efficiently, and its seeds are used in the production of hemp oil, leaving hempseed meal (HSM) as a byproduct, which is reported to be rich in crude protein (CP) around 30–38% on a dry matter basis and fiber, making it a possible feedstuff and a protein source for ruminants. However, limited work has been carried out to evaluate the effects of utilizing HSM as feedstuffs for goats on their carcass characteristics. This study aims to investigate the effects of feeding various levels of HSM on the carcass traits of the crossbred Boer goats. Results suggest that including up to 30% of HSM in the diet of growing meat goats has no adverse effects on their carcass traits and meat quality. These results might be encouraging for the hemp industry as HSM could potentially be marketed and used as an alternative protein source for livestock.

Abstract: Hempseed meal (HSM) is the byproduct of hemp seeds and is rich in crude protein and fiber, making it an ideal candidate as a feedstuff for ruminants. The objective of the present study is to evaluate the effects of feeding different levels of HSM on the carcass traits of crossbred Boer goats. Forty castrated goat kids (approximately six months, 25.63 ± 0.33 kg) were assigned to one of four treatments ($n = 10$) in a completely randomized design. Goats were fed pelleted diets (50% forage and 50% concentrate) with additional supplementation of HSM: control with 0%, 10%, 20%, and 30% of the total diets. Goats were harvested and processed after a 60-day feeding trial. There were no significant differences ($p > 0.05$) in the mean values of dressing percentages, carcass weights, body wall thickness, and ribeye area among treatments. Marbling scores and percentages of moisture, fats, proteins, and collagen in the muscles showed no significant differences ($p > 0.05$) among the treatments. Results suggest that including up to 30% of HSM in the diet of growing meat goats does not affect their carcass traits.

Keywords: carcass traits; goats; hempseed meal; marbling scores; ribeye area

Citation: Gurung, R.; Ale, K.B.; Abrahamsen, F.W.; Moyer, K.; Sawyer, J.T.; Gurung, N.K. Carcass Traits of Growing Meat Goats Fed Different Levels of Hempseed Meal. *Animals* **2022**, *12*, 1986. https://doi.org/10.3390/ani12151986

Academic Editors: Panagiotis Simitzis and Athanasios I. Gelasakis

Received: 19 July 2022
Accepted: 2 August 2022
Published: 5 August 2022

1. Introduction

Industrial hemp (*Cannabis sativa* L.) is an annual herbaceous plant grown mainly for fibers, seeds, and various industrial products [1]. This variety of *Cannabis sativa* is legalized to cultivate as it produces 0.3% or less tetrahydrocannabinol (THC) [2]. However, other varieties of Cannabis, such as marijuana, contains 5–25% and shows more intoxicating and hallucinogenic properties [2]. Industrial hemp is currently being investigated as a potential new crop for livestock in the U.S. with the passage of the 2014 and 2018 Farm Bills in the USA [3,4] and legalized hemp production and cannabidiol (CBD) products derived from hemp [4]. Hemp plants thrive well, and its seeds are utilized in the production of hemp oil. The remaining byproduct is used in the production of hempseed meal (HSM), which is found to be rich in crude protein (CP), approximately 30 to 38% on a dry matter (DM) basis and fiber, making it an ideal candidate as a feedstuff for ruminant animals [1,5–8]. Hempseed oil is also abundant in essential fatty acids. It contains 50 to 70% linoleic acids and 15 to 25% alpha-linolenic acids [9]. It contains around 80% polyunsaturated fatty acids (PUFAs) and essential amino acids, especially arginine [5]. These polyunsaturated essential fatty acids, and easily digestible complete protein properties of hempseeds provide nutritional benefits to humans and animals [10]. It has an ideal omega 6 to omega 3 essential fatty acids ratio for optimal human health [5,11]. These polyunsaturated essential fatty acids can be utilized as another energy source for the animal while potentially improving the immune function [1]. On a DM basis, hempseed meal contains 32.08% CP, 50.79% neutral detergent fiber (NDF), 39.04% acid detergent fiber (ADF), 8.24% ash, and 5.24% ether extract [1]. Studies showed that HSM is an excellent source of rumen undegraded protein [1,5]. Gibb et al. [12] found no detrimental effect on the carcass traits of steers fed 14% full-fat hemp seed (HS). Hessle et al. [6] fed steers diets supplemented with cold-pressed hempseed cake and found no effect on the carcass traits compared to the soybean meal fed steers. There are few studies on the hempseed and its byproduct as a potential protein source for ruminants, especially goats [1,7,12].

Small ruminants such as goats can be vital in production arenas where nutritional resources are limited. Goats (*Capra hircus*) are small ruminants domesticated for meat, milk, fiber, and skins [13]. The meat goat industry is rapidly growing, and its demand and popularity is increasing in the US with the increase in the ethnic populations and immigrants from Asia, Africa, Latin America, and the Middle East [13]. Goat meat is considered healthier than other red meats [14] because it has leaner protein, less fat, and is high in iron and vitamin B12. It also has balanced amino acids, saturated/unsaturated fatty acids, low n6:n3 ratio, and high conjugated linoleic acid [15]. Carcass characteristics are important aspects of meat quality that help to determine the marketing value of the meats and live animals. So, it is highly significant to the producer and consumer. Meat goat carcass characteristics can be influenced by breed, age, sex, diet, and environment [16]. Numerous studies have been conducted to evaluate the effect of various agricultural byproducts on the different carcass characteristics of small ruminants [17–19]. Previous research evaluating the effect of HSM on fresh and cooked characteristics of meat are extremely limited. However, Smith et al. [20] reported no significant difference ($p > 0.05$) on the fresh and cooked characteristics in the goat meat fed with varying levels of HSM. Limited research has been undertaken on the carcass characteristics of goat meat [21]. So, the objective of the current study was to investigate the effects of feeding various levels of HSM on the carcass traits and the meat quality of crossbred Boer goats during the 60 days trial period.

2. Materials and Methods

2.1. Experimental Animals and Diets

The study was conducted at the Caprine Research and Education Unit of George Washington Carver Agricultural Experiment Station of Tuskegee University, Tuskegee, AL, USA. All animal handling, care, and sample collection procedures were conducted and approved by Tuskegee University Animal Care and Use Committee protocol number R07-2019-5. Goats were brought from Texas and quarantined for 14 days during which they were provided control complete total mixed ration diet. The goats were individually housed in 1.1 × 1.2 m pens with plastic-coated expanded metal floors. The goats were vaccinated with *Clostridium perfringens* type C and D-Tetani Bacterin-Toxoid (Bayer Corp., Shawnee Mission, KS, USA) and dewormed with Cydectin (moxidectin; Fort Dodge Animal Health, Fort Dodge, IA, USA) before arrival. Forty castrated Boer cross goats *(Capra aegagrus hircus)* with approximately six months of age and an initial average weight of 25.63 ± 0.33 kg were randomly allocated to one of four treatments ($n = 10$) in a completely randomized design for 60 days. Treatments consisted of different levels of HSM: control with 0%, 10%, 20%, and 30% HSM supplementation of the total diets.

Hempseed meal was obtained from Kentucky Hemp Works, Crofton, KY, USA. Pelleted diets were prepared at Auburn University Poultry Feed Mill to ensure the goats consume as much HSM as possible by reducing the chance of selective feeding by goats. The complete diet consisted of bermudagrass (*Cynodon dactylon*) hay, soybean meal, meat maker 16:8 (goat premix), cracked corn, molasses, and HSM at varying rates. Goat diets consist of 50% concentrate mix (as-fed basis) and 50% mixed hay which were offered separately. The goats were fed pelleted diets formulated to meet their nutritional needs. Animals were fed twice a day with ad libitum access to water throughout the experiment. Additionally, each animal was provided with 0.23 kg of mixed hay to aid in optimal rumen function. Feed remnants were weighed twice daily and replaced to ensure animals had constant access to fresh feed.

2.2. Chemical Analysis

The samples of hay and concentrate mixes were separated and collected before the experiment and dried for 48 hours at 55 °C in a convection oven (model 420, NAPCO, Pittsburgh, PA, USA). The samples were ground using a grinding machine (Hammer Mill Model 1250; Lorenz MFG Co., Benson, MN, USA) and sent to the Holmes laboratories, Millersburg, Ohio for analysis. Dry matter, crude protein, acid detergent fiber, crude fat, ash, calcium, and phosphorus were completed according to the methods described by the American Organization of Analytical Chemists [22]. The D.M. concentration of hay was measured at 60 °C for 24 hours, and ash was determined at a temperature of 550 °C for 5 hours. Nitrogen concentration (N) of the diet samples was determined using Kjeldahl N method, and CP was measured by: $Crude\ Protein\ (CP) = Nitrogen\ (N) \times 6.25$

Neutral detergent fiber (NDF) and acid detergent fiber (ADF) were calculated using the method of van Soest et al. [23] as utilized by the Ankom Fiber Analyzer (Ankom Technology Corp., Macedonia, NY, USA). Lignin concentration was determined according to methods described by the United States Department of Agriculture [24].

2.3. Slaughter and Carcass Evaluation

On day 60 of the study, goats were weighed (final weight) utilizing a goat and sheep scale manufactured by Lakeland Farm and Ranch (Clawson, MI, USA) in 0.1 kg increments. Feed and water were withheld overnight before slaughter. The goats were transported and harvested according to the USDA standards at the Lambert-Powell Meats Laboratory, Auburn University, AL, USA. Goats were slaughtered using approved methods of US Department of Agriculture-Food Safety and Inspection Service (USDA FSIS) according to the Humane Slaughter Act [25]. After the slaughter process, the carcass was rinsed with hot water and a 2% solution of lactic acid and weighed using a Static Monorail Scale (Vandenberg Scales, Sioux Center, IA, USA) to determine the hot carcass weight (HCW). Kidney, pelvic, and heart (KPH) were collected to determine KPH fat percentage of the carcass weight using an analytical balance (PB3002-S, Met- tler Toledo, Columbus, OH, USA). Carcasses were then chilled at 4 °C for 24 hours, after chilling, carcasses were re-weighed using the Static Monorail Scale to determine the cold carcass weight (CCW). Dressing percentage (DP) was calculated by the following equation:

$$Dressing\ Percentage\ (DP) = \left(\frac{Hot\ Carcass\ Weight}{Live\ Weight}\right) \times 100$$

Then, Ribeye area (REA) was determined by measuring the surface area of the longissimus dorsi muscle between the 12th and 13th ribs of the goat carcass using a grid [26]. Marbling scores were measured according to the USDA [27] utilizing beef marbling scores as reference, between the 12th and 13th rib interface. Fat thickness opposite to the loin eye was determined by measuring the subcutaneous fat over the ribeye area (REA) utilizing a caliper after measuring the REA. In addition, body wall thickness was measured approximately 4.5 inches from the midline of the ribbed goat carcass utilizing a stainless-steel ruler. Fresh meat samples were minced twice through a meat-grinding machine; minced meat samples were vacuum packaged and frozen for chemical analysis. The chemical composition of meat samples was determined by utilizing the proximate analysis to determine the moisture, fat, protein, and collagen content according to the Association of Official Analytical Chemists [28]. Samples for chemical analysis (protein, moisture, fat, and collagen) were conducted using a near-infrared (NIR) approved spectrophotometer (Food Scan™, FOSS Analytical A/S, Hilleroed, Denmark).

2.4. Statistical Analysis

Data were analyzed using the GLIMMIX procedures of SAS 9.4 (SAS Inst. Inc., Cary, NC, USA). Treatment served as the lone fixed effect for carcass measurements and physiochemical components. Least squares means were generated and statistical significance ($p < 0.05$), F-values were observed, and least squares means were separated using pair-wise t-tests (PDIFF option).

3. Results and Discussion

3.1. Diet Composition

The chemical composition of Bermuda grass hay (BGH) and hempseed meal (HSM) are shown in Table 1, while the ingredients and chemical composition of concentrate mixes are shown in Table 2. HSM is relatively high in CP concentration in the present study, and other studies reported similar values [1,6–8,29]. In this study, diets were balanced to be iso-nitrogenous replicating conditions that might be experienced in a production scenario. As the rate of HSM supplementation increased, there was a decrease in total digestible nutrients (TDN) and non-fiber carbohydrate (NFC), while there was an increase in crude fat. Fiber content, both ADF and NDF increased as the rate of supplementation increased. Lignin concentration, the indigestible portion of plant material, also increased with the increasing level of HSM. Phosphorus concentration also increased with an increasing level of supplementation while calcium decreased. As the level of HSM increased, net energy for gain (NEg) and TDN decreased, which could potentially induce a difference in growth and carcass characteristics; however, no significant differences were observed among treatments.

Table 1. Chemical composition of Bermuda grass hay (BGH) and hempseed meal (HSM) used in the experiment.

[a] Items	Diet	
	BGH	HSM
Dry Matter (%)	95.85	89.61
Crude Protein (%)	9.58	36.42
Acid detergent fiber (%)	42.82	36.47
Neutral detergent fiber (%)	70.99	49.47
Lignin (%)	6.68	12.76
Crude Fat (%)	1.23	11.53
Total digestible nutrients (%)	55.13	63.22
Net energy for gain (Mcal/lb)	0.266	0.377
Ash (%)	4.79	5.82
Calcium (%)	0.26	0.23
Phosphorus (%)	0.17	1.03

[a] Values are presented on a dry matter basis, except dry matter. Source: Holmes Laboratory Inc., Millersburg, OH 44654, USA.

Table 2. Ingredients and nutrient composition of the experimental diets fed to Boer crossbred meat goats.

Items [a]	Diets			
	0% HSM	10% HSM	20% HSM	30% HSM
Ingredients, as-fed basis				
Bermuda Grass Hay (%)	50	50	50	50
Cracked Corn (%)	28	24	18.5	14
Soybean Meal (%)	18.5	13	8	2.5
Molasses (%)	2.5	2.5	2.5	2.5
[b] Goat Premix (%)	1	1	1	1
Nutrient composition, DM basis				
Dry Matter (%)	89.06	88.66	89.1	89.86
Crude Protein (%)	19.18	19.91	19.25	20.39
Lignin (%)	3.34	4.77	6.21	7.02
Acid Detergent Fiber (%)	21.06	24.67	28.96	30.97
Neutral Detergent Fiber (%)	33.29	35.18	39.63	42.8
Non-Fiber Carbohydrate (%)	40.64	37.8	35.38	31.9
Acid Hydrolysis Fat (%)	3.19	3.32	4.22	4.5
Total Digestible Nutrients (%)	71.2	69.2	64.7	62.79
Net Energy for gain (Mcal/lb)	0.481	0.456	0.397	0.372
Ash (%)	7.02	7.01	7.09	6.79
Calcium (%)	0.95	0.92	0.88	0.82
Phosphorus (%)	0.39	0.41	0.48	0.52

[a] Values are presented on a dry matter basis, except DM. [b] Goat premix (%): Ca 9.0, P 8.0, salt 41.0, K 0.1, Mg 1.0; (ppm) Cu 1750, Se 25.0, Zn 7500 and (IU kg^{-1}) Vitamin A 3,08,644, Vitamin D 24,251, and Vitamin E 1653.

3.2. Carcass Characteristics

The effect of different levels of HSM supplementation on carcass traits of goats are presented in Tables 3 and 4. HSM supplementation showed no significant effect on the carcass weights ($p > 0.05$) evaluated in the present study (Table 3). There were no significant differences ($p > 0.05$) in the mean values of dressing percentages (DP) among treatments (Table 3). The values were 46.59, 45.42, 45.77, and 46.16% for diets containing 0, 10, 20, and 30% HSM, respectively, which were higher than that reported by Gurung et al. [18] when Dried Distillers Grain was used with soluble (DDGS) for goats. However, the DP was lower than that obtained when goats were supplemented with peanut skins [30], and grain diets and pasture [31]. The values of hot carcass weight (HCW) were 16.55, 16.37, 15.69, and 15.33 kg for diets containing 0, 10, 20, and 30% HSM, respectively, which are in decreasing order with increase of HSM supplementation but not significantly different ($p > 0.05$) among the treatments (Table 3). Similarly, the values of cold carcass weight (CCW) of the animals were decreasing with the increase in the levels of HSM supplementation but were not significantly different ($p > 0.05$) among the treatments (Table 3). The values of HCW and CCW of meat goats were higher than that reported by Ebrahimi et al. [32], whereas they were lower than that reported by Min et al. [30].

The marbling scores (376, 399, 355, 364, respectively) were also not significantly different ($p > 0.05$) among the treatments and fall on the marbling degree of "Slight (S.L.)". Flank color and marbling scores were all within the normal range for growing meat goats, as these values are consistent with other values reported by other researchers [17–19,30,32]. Ribeye area (REA) was not significantly different ($p > 0.05$) among treatments with values of 3.68, 3.4, 3.47, 3.39 cm^2 for 0, 10, 20, and 30% HSM, respectively. There were no differences in the body wall (BW) thickness ($p > 0.05$) among treatments (Table 3), with 0.31, 0.30, 0.29, and 0.29 cm for 0, 10, 20, and 30 % HSM, respectively. The results were lower than those of Gurung et al. 2009 [18], reported the BW thickness of 1.09 cm for goats fed with 10.3 % DDGS and 3.9 cm BWF thickness reported by 50 % peanut skins (PS) supplementation on the goats [30]. Similarly, back fat thickness measured opposite loin eye was not significantly different ($p > 0.05$) among treatments. The values were 0.033, 0.027, 0.032, and 0.025 cm for diets containing 0, 10, 20, and 30% HSM, respectively. The results were lower than the result of 0.8 cm of fat depth for 30% PS supplementation presented by Min et al. [30]. The KPH fat percentage was not affected ($p > 0.05$) by varying levels of HSM supplementation in the meat goats. These results suggest that varying levels of HSM supplementation have no negative impact on the carcass characteristics of goat meat.

Table 3. Effects of varying levels of HSM supplementation on the carcass yields and quality measurements of meat goats.

Items	Treatment, %				SEM	*p*-Value
	0	10	20	30		
Dressing Percent, %	46.59	45.43	45.77	46.17	0.510	0.107
Hot carcass weight, kg	16.55	16.37	15.69	15.33	0.42	0.153
Cold carcass weight, kg	16.42	16.09	15.51	15.15	0.418	0.152
Ribeye Area, cm^2	3.68	3.40	3.47	3.39	0.119	0.297
Fat Thickness Opposite Loin Eye, cm	0.033	0.027	0.032	0.025	0.004	0.556
Body Wall Thickness, cm	0.31	0.30	0.29	0.29	0.036	0.975
Kidney, Pelvic, Heart Fat, %	2.20	2.70	2.60	2.45	0.295	0.656
Marbling score *	376	399	355	364	16.5	0.278

* Means based on USDA (2001) marbling scores where 100 = practically devoid (Standard$^-$), 200 = traces (Standard$^{+/0}$), 300 = slight (Select$^{+/-}$), 400 = small (Choice$^-$), 500 = modest (Choice^0), 600 = moderate (Choice$^+$), 700 = slightly abundant (Prime$^-$) and 800 = moderately abundant (Prime^0). SEM = standard error of mean.

Table 4. Physio-chemical attributes of goat meats supplemented with varying levels of HSM.

Meat Quality Traits	Treatment, %				SEM	*p*-Value
	0	10	20	30		
Moisture, %	71.28	72.82	71.24	73.58	0.757	0.087
Fat, %	13.69	11.49	13.68	10.18	1.069	0.065
Protein, %	22.95	22.96	23.09	23.72	0.383	0.446
Collagen, %	2.53	2.31	2.52	2.29	0.075	0.056

SEM = standard error of mean.

The results reported by Gurung et al. [18] also agreed with the results of the present study that varying levels of agricultural protein by products do not significantly impact carcass characteristics. Furthermore, Abdelrahim et al. [17] fed varying levels of DDGS (12.7 and 25.4% of diet) to growing lambs and determined that the different levels of DDGS had no negative impact on carcass characteristics evaluated in that study. There was no significant difference ($p > 0.05$) in the moisture, fat, protein, and collagen content of the goat meats among treatments, as shown in Table 4. The percentage of moisture, fat, protein, and collagen were all within the normal range for growing meat goats, consistent with the values reported by other studies [17–19,30,32]. Similarly, moisture and protein contents were consistent with the values reported by Sen et al. and Shija et al. [31,33]. However, the fat percentage of goat meat was higher than that reported by previous researchers [31,33]. These results showed that including up to 30% HSM supplementation has no detrimental effects on the physio-chemical attributes of goat meat.

4. Conclusions

Limited studies have been implemented regarding the effect of supplementing HSM on the carcass traits and meat quality of growing meat goats. So, these results suggest that up to 30% HSM can be fed to growing meat goats without affecting the carcass traits. HSM could potentially be an effective protein source for growing meat goats since even the 30% level has no adverse effects. These results might be promising for the future of the industrial hemp industry as HSM could potentially be marketed as an alternative protein source for livestock. Further works are needed to evaluate the effects of HSM inclusion rates on the carcass traits and meat quality of goats.

Author Contributions: Conceptualization, N.K.G., J.T.S.; methodology, J.T.S.; software, J.T.S.; validation, N.K.G., F.W.A.; formal analysis, J.T.S.; investigation, F.W.A., R.G., K.B.A.; resources, K.M., J.T.S., N.K.G.; data curation, F.W.A., J.T.S.; writing—original draft preparation, R.G.; writing—review and editing, R.G., F.W.A., K.B.A., J.T.S.; visualization, N.K.G., J.T.S.; supervision, J.T.S., N.K.G.; project administration, N.K.G. All authors have read and agreed to the published version of the manuscript.

Funding: This research was funded by USDA/NIFA Capacity Building Grant (Award #: 2021-38821-34716).

Institutional Review Board Statement: Goats used for the current study were cared for under Tuskegee University IACUC (TUACUC) protocol request number: R07-2019-5. All methods of animal handling and care were approved prior to commencing research activities.

Data Availability Statement: Data available upon reasonable request from the corresponding author.

Acknowledgments: Appreciation is extended to the George Washington Carver Agricultural Experiment Station, Tuskegee University and Lambert-Powell Meat Laboratory, Auburn University with procurement of raw material and processing technologies. Furthermore, the authors are grateful for the support provided by Kentucky hemp works for providing hempseed meals.

Conflicts of Interest: The authors declare no conflict of interest.

References

1. Mustafa, A.F.; McKinnon, J.J.; Christensen, D.A. The Nutritive Value of Hemp Meal for Ruminants. *Can. J. Anim. Sci.* **1999**, *79*, 91–95. [CrossRef]
2. Small, E.; Marcus, D. Tetrahydrocannabinol Levels in Hemp (Cannabis Sativa) Germplasm Resources. *Econ. Bot.* **2003**, *57*, 545–558. [CrossRef]
3. Cherney, J.; Small, E. Industrial Hemp in North America: Production, Politics and Potential. *Agronomy* **2016**, *6*, 58. [CrossRef]
4. Johnson, R. Hemp as an Agricultural Commodity. *Congr. Res. Serv.* **2018**, *7*, 5700.
5. Callaway, J.C. Hempseed as a Nutritional Resource: An Overview. *Euphytica* **2004**, *140*, 65–72. [CrossRef]
6. Hessle, A.; Eriksson, M.; Nadeau, E.; Turner, T.; Johansson, B. Cold-Pressed Hempseed Cake as a Protein Feed for Growing Cattle. *Acta Agric. Scand. Sect. A—Anim. Sci.* **2008**, *58*, 136–145. [CrossRef]
7. Karlsson, L.; Finell, M.; Martinsson, K. Effects of Increasing Amounts of Hempseed Cake in the Diet of Dairy Cows on the Production and Composition of Milk. *Animal* **2010**, *4*, 1854–1860. [CrossRef]
8. Karlsson, L.; Ruiz-Moreno, M.; Stern, M.D.; Martinsson, K. Effects of Temperature during Moist Heat Treatment on Ruminal Degradability and Intestinal Digestibility of Protein and Amino Acids in Hempseed Cake. *Asian-Australas. J. Anim. Sci.* **2012**, *25*, 1559–1567. [CrossRef]
9. Leizer, C.; Ribnicky, D.; Poulev, A.; Dushenkov, S.; Raskin, I. The Composition of Hemp Seed Oil and Its Potential as an Important Source of Nutrition. *J. Nutraceuticals Funct. Med. Foods* **2000**, *2*, 35–53. [CrossRef]
10. Deferne, J.L.; Pate, D.W. Hemp Seed Oil: A Source of Valuable Essential Fatty Acids. *J. Int. Hemp Assoc.* **1996**, *3*, 1–7.
11. Oomah, B.D.; Busson, M.; Godfrey, D.V.; Drover, J.C.G. Characteristics of Hemp (*Cannabis Sativa* L.) Seed Oil. *Food Chem.* **2002**, *76*, 33–43. [CrossRef]
12. Gibb, D.J.; Shah, M.A.; Mir, P.S.; Mcallister, T.A.; Mir, M.A.; Mcallister, P.S. Effect of Full-Fat Hemp Seed on Performance and Tissue Fatty Acids of Feedlot Cattle. *Can. J. Anim. Sci.* **2005**, *85*, 223–230. [CrossRef]
13. Osti, S.; Gillespie, J.; Nyaupane, N.P.; McMillan, K. Meat Goat Production in the United States: Adoption of Technologies, Management Practices, and Production Systems. *J. ASFMRA* **2016**, *2015*, 116–129.
14. Anaeto, M.; Adeyeye, J.; Chioma, G.; Olarinmoye, A.; Tayo, G. Goat Products: Meeting the Challenges of Human Health and Nutrition. *Agric. Biol. J. N. Am.* **2010**, *1*, 1231–1236. [CrossRef]
15. Chin, S.F.; Liu, W.; Storkson, J.M.; Ha, Y.L.; Pariza, M.W. Dietary Sources of Conjugated Dienoic Isomers of Linoleic Acid, a Newly Recognized Class of Anticarcinogens. *J. Food Compos. Anal.* **1992**, *5*, 185–197. [CrossRef]
16. Irshad, A.; Gurunathan, K.; Kumar, S.; Kumar, A.; Kumar, A.; MR, V.; Shukla, V. Factors Influencing Carcass Composition of Livestock: A Review. *J. Anim. Prod. Adv.* **2013**, *3*, 177. [CrossRef]
17. Abdelrahim, G.M.; Khatiwada, J.; Gurung, N.K. Effects of Dried Distillers Grains with Solubles on Performance and Carcass Characteristics of Lamb. *J. Anim. Res. Technol.* **2017**, *1*, 25–30. [CrossRef]
18. Gurung, N.K.; Solaiman, S.G.; Rankins, D.L.; McElhenney, W.H. Effects of Distillers Dried Grains with Solubles on Feed Intake, Growth Performance, Gain Efficiency and Carcass Quality of Growing Kiko x Spanish Male Goats. *J. Anim. Vet. Adv.* **2009**, *8*, 2087–2093.
19. Hatamleh, S.M.; Obeidat, B.S. Growth Performance and Carcass Traits Responses to Dried Distillers' Grain with Solubles Feeding of Growing Awassi Ram Lambs. *Animals* **2019**, *9*, 954. [CrossRef]
20. Smith, H.R.; Abrahamsen, F.W.; Rehm, J.G.; Wilborn, B.; Blythe, E.; Sawyer, J.T.; Gurung, N. PSV-31 Influence of Hempseed Meal Supplementation on Fresh and Cooked Characteristics of Boer Cross Goats. *J. Anim. Sci.* **2020**, *98*, 362. [CrossRef]

21. Asif Arain, M.; Khaskheli, M.; Rajput, I.R.; Faraz, S.; Rao, S.; Umer, M.; Devrajani, K. Effect of Slaughtering Age on Chemical Composition of Goat Meat. *Pak. J. Nutr.* **2010**, *9*, 404–408. [CrossRef]
22. AOAC. *Official Methods of Analysis*, 15th ed.; American Organization of Analytical Chemists: Arlington, VA, USA, 1990.
23. Van Soest, P.J.; Robertson, J.B.; Lewis, B.A. Methods for Dietary Fiber, Neutral Detergent Fiber, and Nonstarch Polysaccharides in Relation to Animal Nutrition. *J. Dairy Sci.* **1991**, *74*, 3583–3597. [CrossRef]
24. USDA. Forage Fiber Analyses (Apparatus, Reagents, Procedures, and Some Applications). *USDA Agric. Handb.* **1970**, *379*, 9–11.
25. USDA. *Humane Handling of Livestock and Poultry*; Food Safety and Inspection Service, United States Department of Agriculture: Washington, DC, USA, 2015.
26. Yáñez, E.A.; Ferreira, A.C.D.; Medeiros, A.N.; Pereira Filho, J.M.; Teixeira, I.A.M.A.; Resende, K.T. Methodologies for Ribeye Area Determination in Goats. *Small Rumin. Res.* **2006**, *66*, 197–200. [CrossRef]
27. USDA. *Institutional Meat Purchased Specifications for Fresh Goat*; Series 11; Meat Grading Certification Branch; USDA, MRP, AMF, Livestock and Seed Program: Washington, DC, USA, 2001.
28. AOAC. *Official Methods of Analysis*, 18th ed.; Association of Official Analytical Chemists: Arlington, VA, USA, 2007.
29. Iannaccone, M.; Ianni, A.; Contaldi, F.; Esposito, S.; Martino, C.; Bennato, F.; de Angelis, E.; Grotta, L.; Pomilio, F.; Giansante, D.; et al. Whole Blood Transcriptome Analysis in Ewes Fed with Hemp Seed Supplemented Diet. *Sci. Rep.* **2019**, *9*, 16192. [CrossRef]
30. Min, B.R.; Frank, A.; Gurung, N.; Lee, J.H.; Joo, J.W.; Pacheco, W. Peanut Skin in Diet Alters Average Daily Gain, Ruminal and Blood Metabolites, and Carcass Traits Associated with Haemonchus Contortus Infection in Meat Goats. *Anim. Nutr.* **2019**, *5*, 278–285. [CrossRef] [PubMed]
31. Sen, A.R.; Santra, A.; Karim, S.A. Carcass Yield, Composition and Meat Quality Attributes of Sheep and Goat under Semiarid Conditions. *Meat Sci.* **2004**, *66*, 757–763. [CrossRef]
32. Ebrahimi, M.; Rajion, M.A.; Jafari, S.; Faseleh Jahromi, M.; Oskoueian, E.; Qurni Sazili, A.; Goh, Y.M.; Ghaffari, M.H. Effects of Dietary N-6: N-3 Polyunsaturated Fatty Acid Ratios on Meat Quality, Carcass Characteristics, Tissue Fatty Acid Profiles, and Expression of Lipogenic Genes in Growing Goats. *PLoS ONE* **2018**, *13*, e0188369. [CrossRef]
33. Shija, D.S.; Mtenga, L.A.; Kimambo, A.E.; Laswai, G.H.; Mushi, D.E.; Mgheni, D.M.; Mwilawa, A.J.; Shirima, E.J.M.; Safari, J.G. Chemical Composition and Meat Quality Attributes of Indigenous Sheep and Goats from Traditional Production System in Tanzania. *Asian-Australas. J. Anim. Sci.* **2013**, *26*, 295–302. [CrossRef]

animals

Article

Enrichment of Brain n-3 Docosapentaenoic Acid (DPA) and Retinal n-3 Eicosapentaenoic Acid (EPA) in Lambs Fed *Nannochloropsis oceanica* Microalga

Ana C. M. Vítor [1,2,3,*], Jorge J. Correia [1,2,3], Susana P. Alves [1,2,3] and Rui J. B. Bessa [1,2,3]

1 Faculdade de Medicina Veterinária, Universidade de Lisboa, Avenida da Universidade Técnica, 1300-477 Lisboa, Portugal
2 CIISA—Centro de Investigação Interdisciplinar em Sanidade Animal, Avenida da Universidade Técnica, 1300-477 Lisboa, Portugal
3 Associate Laboratory for Animal and Veterinary Sciences (AL4AnimalS), Avenida da Universidade Técnica, 1300-477 Lisboa, Portugal
* Correspondence: anacrisvitor@fmv.ulisboa.pt

Simple Summary: Omega-3 polyunsaturated fatty acids (n-3 PUFAs), mainly eicosapentaenoic acid (EPA) and docosahexaenoic acid (DHA), have been increasingly studied due to their beneficial health effects. N-3 PUFAs are particularly abundant in the brain and retina, where they play various roles that are important to the maintenance of normal function in those organs. The present study aimed to evaluate the FA profile of lamb brain and retinal tissues after they were fed three experimental diets supplemented with an EPA-rich microalga for 21 days. The microalga was delivered in a different format in each one of the diets (oil, spray-dried and freeze-dried biomass); therefore, its efficiency in altering the FA profile of brain and retina was evaluated for each diet. Overall, our results demonstrated that the brain EPA content remained unchanged after EPA supplementation, in contrast with the retinal EPA, which was very responsive to microalga supplementation.

Abstract: Omega-3 polyunsaturated fatty acids (n-3 PUFAs) have special physiological functions in both brain and retinal tissues that are related to the modulation of inflammatory processes and direct effects on neuronal membrane fluidity, impacting mental and visual health. Among them, the long-chain (LC) n-3 PUFAs, as eicosapentaenoic acid (EPA) and docosahexaenoic acid (DHA), are of special importance. Scarce data are available about the fatty acid (FA) composition of the ruminant brain in response to dietary intervention. However, we decided to examine the brain and retina FA composition of lambs supplemented with an EPA-rich microalga feed for 21 days, as it is known that despite the extensive biohydrogenation of dietary PUFAs in the rumen, ruminants can selectively accumulate some n-3 LC-PUFAs in their brain and retinal tissues. Twenty-eight male lambs were fed a control diet, or the same diet further supplemented with *Nannochloropsis* sp. microalga. Their brains and retina were collected for FA characterization. Overall, the brain FA profile remained unchanged, with little alteration in omega-3 docosapentaenoic acid (DPA) enhancement in both the hippocampus and prefrontal cortex. Retinal tissues were particularly responsive to the dietary intervention, with a 4.5-fold enhancement of EPA in the freeze-dried-fed lambs compared with the control lambs. We conclude that retinal tissues are sensitive to short-term n-3 PUFA supplementation in lambs.

Keywords: fatty acid; hippocampus; prefrontal cortex; dimethyl acetal

Citation: Vítor, A.C.M.; Correia, J.J.; Alves, S.P.; Bessa, R.J.B. Enrichment of Brain n-3 Docosapentaenoic Acid (DPA) and Retinal n-3 Eicosapentaenoic Acid (EPA) in Lambs Fed *Nannochloropsis oceanica* Microalga. *Animals* **2023**, *13*, 828. https://doi.org/10.3390/ani13050828

Academic Editors: Panagiotis Simitzis and Athanasios I. Gelasakis

Received: 20 January 2023
Revised: 20 February 2023
Accepted: 21 February 2023
Published: 24 February 2023

1. Introduction

Fatty acids (FAs) are the most abundant organic compounds in the brain. More than 90% of polyunsaturated FAs (PUFAs) in the mammalian brain are composed of long-chain PUFAs (i.e., $\geq$20 C chain, LC-PUFA) such as arachidonic acid (AA, 20:4n-6) and docosahexaenoic acid (DHA, 22:6n-3) [1]. Most of these PUFAs are esterified in brain membrane phospholipids (PLs), and they influence brain functions by altering the biophysical properties of cell membranes [2]. Omega-3 PUFAs (n-3 PUFAs) have a special importance since they play a role in a variety of physiological functions related to neurogenesis, neurotransmission, and neuroinflammation, contributing to the development, functioning, and ageing of the brain. Furthermore, in humans, n-3 LC-PUFA dietary deficiencies are associated with an increased risk of developing various psychiatric disorders [3] and neurodegenerative diseases [4]. Among these FAs, eicosapentaenoic acid (EPA, 20:5n-3) and DHA have been linked to the maintenance of mental health, mediated by the modulation of inflammatory processes and direct effects on neuronal membrane fluidity and receptor function [3]. N-3 docosapentaenoic acid (DPA, 22:5n-3) also plays an important role, as it is the second most abundant n-3 LC-PUFA in the brain after DHA. It is suggested to be specifically beneficial for elderly neuroprotection and early-life brain development [5].

In addition to the brain, retinal tissue also is rich in lipids, which comprise approximately 20% of the retina's dry weight. Retinal membrane PLs have the highest level of LC-PUFAs of any tissue in humans (approximately 33%) [6–8]. The retina is a tissue with a naturally high content of n-3, particularly DHA, which plays an essential role in optimizing the fluidity of photoreceptor membranes, retinal integrity and visual function [9]. DHA also has a protective role in the retina, participating in the anti-inflammatory activity, anti-angiogenesis, anti-apoptosis and providing protection from neurotoxicity [9].

The brain lipids of ruminant species, especially cattle, have been well characterized and appear to be very similar to those of the human brain in terms of both content and composition [10]. Although they are less well-characterized, the lipids in the brains of sheep also do not seem to differ markedly from the ones found in cattle [10]. In ruminants, brain characterization has been traditionally focused on the types of lipids and less so on a deeper characterization of FAs. Moreover, lipid analyses have been performed primarily in brain homogenates or gross anatomical structures within the brain [10], not in specific functional regions or tracts of the ruminant brain. Bovine and ovine retinal fatty acid compositions are similar: they are composed of appreciable amounts of palmitic acid (16:0), stearic acid (18:0), DHA and AA [10,11]. Thus, despite the extensive biohydrogenation of dietary PUFAs in the rumen, ruminants can selectively accumulate n-3 LC-PUFA in their brain and retinal tissues, contrasting with the low deposition of these FAs in adipose tissue and muscle [12]. Liver stores of n-3 LC-FA were reported to be the primary source of these FAs for the brain tissues of rats [13], even during periods of low dietary intake of these FAs [13,14]. This makes it possible to maintain adequate levels of DHA and AA in the brains of DHA- and AA-deprived animals [13,14]. In a very recent publication, cattle supplemented with calcium salts of fish oil with 11% EPA and 8% DHA had, across most brain regions, greater EPA concentrations when compared to palm-oil-supplemented animals [15].

The dietary supplementation of microalgae has been shown to enhance n-3 LC-PUFAs along lambs' gastrointestinal tracts [16]. Moreover, Vítor et al. [16] showed that the drying method applied to the microalgae strongly influenced the powder architecture and cell wall integrity, consequently affecting the degree of EPA protection against rumen microbes. Therefore, we hypothesise that lambs' brains and/or retinal tissues would be sensitive to the differences in n-3 LC-PUFA absorption due to the changes in rumen biohydrogenation associated with the processing of microalgae biomass. Thus, we collected brain and retina samples from six-month-old lambs used in a previous experiment and conducted a detailed FA composition of those tissues.

2. Materials and Methods

2.1. Animal Handling and Diets

The current lamb trial was conducted in compliance with the ARRIVE and international guidelines. The trial was conducted in certified facilities and was approved by an ethical and animal well-being commission, as fully detailed in Vítor et al. [16]. Twenty-eight sixty-day-old Merino Branco ram lambs with an average body weight of 21.8 ± 4.4 kg were housed in INIAV facilities in Santarém, Portugal. The animals were randomly allocated to individual pens (1.52 m^2) with ad libitum access to clean water. The lambs were sorted into four experimental groups with seven replicates per group. The experimental diets included a control diet (C diet), consisting of pellets containing dehydrated lucerne, barley and soybean meal and no added sources of EPA, and three diets supplemented with the microalga *Nannochloropsis* sp., designed to provide approximately 3 g of EPA per kg of diet dry matter (DM). The average content of EPA (mg EPA/g product) in each microalgal format was 235 in the *Nannochloropsis* oil, 22.7 in the spray-dried *Nannochloropsis oceanica* and 30.8 in the lyophilized *Nannochloropsis oceanica*. *Nannochloropsis* sp.-containing diets were composed of the C diet plus 123 g/kg of spray-dried *Nannochloropsis oceanica* biomass (SD diet); 92 g/kg freeze-dried *Nannochloropsis oceanica* biomass (FD diet); and 12 g/kg of *Nannochloropsis* sp. free-oil (O diet) (Table 1). The trial had a 3 week duration limitation due to the high cost of the spray-dried *Nannochloropsis oceanica* biomass and the difficulty of obtaining enough freeze-dried biomass with lab-scale equipment.

Table 1. Total fatty acid content (g/kg dry matter) and fatty acids (FAs) profile (% of total fatty acids) of the experimental diets.

Item	Diets [2]			
	C	O	SD	FD
Total FA [1]	13.7	20.6	19.5	20.2
FA profile				
14:0	n.d.	1.26	2.40	2.28
16:0	25.5	23	23.9	26.2
*c*9-16:1	0.51	5.15	10.2	8.22
17:0	0.66	0.34	n.d.	n.d.
18:0	4.09	2.96	2.51	3.32
*c*9-18:1	18.6	14.5	11.7	13.7
*c*11-18:1	0.73	0.93	0.72	0.69
18:2n-6	41	31.5	28.2	29.1
18:3n-3	8.91	6.80	5.02	6.44
20:4n-6	n.d	2.58	3.69	3.17
20:5n-3	n.d.	10.40	11.3	6.88
22:0	n.d.	0.58	0.36	n.d.

[1] FA—fatty acids. [2] C—control diet with no EPA sources; O—diet with *Nannochloropsis* sp. oil; SD—diet with spray-dried *Nannochloropsis oceanica* biomass; FD—diet with freeze-dried *Nannochloropsis oceanica* biomas; n.d.—not detected. In the FA notation (x:n-), 'x' represents the number of C atoms, ':' the number of double bonds and 'n-' the location, in its carbon chain, of the double bond which is closest to the methyl end of the molecule. *c* stands for *cis*. Adapted from [16].

2.2. Slaughter and Sample Collection

After the end of the third week, the animals were slaughtered using a captive bolt. This was followed by exsanguination. The brain tissue was removed whole. It retained its shape and landmarks in spite of the captive bolt damage. The brain was cut on a sagittal plane and divided into two hemispheres, which were then frozen at −80 °C. After thawing, the different parts were individualized from the right hemisphere (Figure 1), stored in individual bags, frozen at −80 °C and lyophilized. Grey and white matter were collected from two different points in the brain and were considered samples from non-function-specific brain parts, representing only samples from the two histologic and physiological brain areas. The prefrontal cortex (cerebral cortex covering the front part of the frontal lobe), and hippocampus (located in the medial part of the temporal lobe and, on a mid-sagittal section of the brain, posterior to the amygdala extending posteriorly to the splenium of the corpus callosum) were selected as function-specific brain parts.

Immediately after slaughter, the right eyeball of each lamb was removed with a spatula, stored in a bag and frozen at −80 °C. The eyeballs were thawed, and the retina and *tapetum lucidum* (RTL) were individualised (Figure 2). The liver was removed from the carcass, and a portion of the left lobe was stored in a bag and frozen at −80 °C. All brain parts, RTL and a portion of the liver were lyophilised prior to fatty acid extraction.

Figure 1. Lamb right brain hemisphere. The four collected parts are highlighted in the figure: prefrontal cortex and hippocampus are filled in blue and yellow, respectively; grey matter is identified with the yellow triangles and white matter with the blue circles.

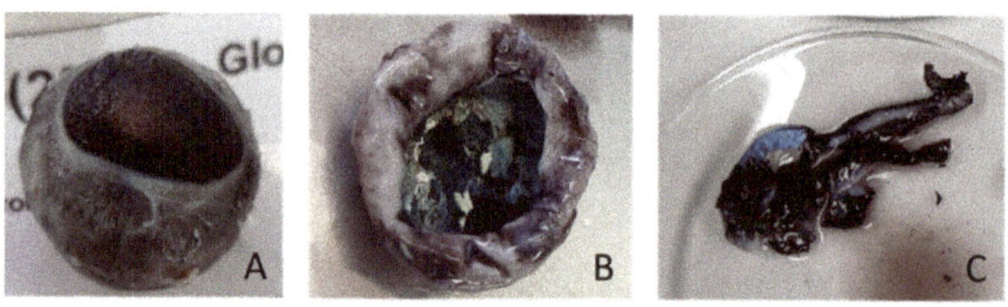

Figure 2. Dissection of the right eyeball of one lamb. From left to right: (**A**) right eyeball; (**B**) removal of eye structures (cornea, iris and lens) and *tapetum lucidum* evidenced; and (**C**) cutting of material used for FA analysis including *tapetum lucidum* and retina tissues.

2.3. Fatty Acid Methyl Esters (FAMEs) and Dimethyl Acetals (DMAs) Analysis

Fatty acid methyl esters (FAMEs) and dimethyl acetals (DMAs) of the brain, RTL tissues and liver samples were prepared by acid-catalysed transesterification in methanol [17]. In plasmalogens, the sn-1 position of the glycerol contains a vinyl-ether that releases a DMA in the presence of acid methanol solution. Briefly, approximately 100 mg of lyophilised and ground sample was weighted to reaction tubes. Toluene and 1 mL of internal standard (methyl nonadecanoate −1 mg/mL) were then added, and the samples were placed in an ultrasound bath for ten minutes. A solution of 1.25 M HCl in methanol (3 mL) was added and left to react overnight at 50 °C in a water bath.

Fatty acid methyl esters and DMAs were analysed by gas chromatography with flame ionization detection (GC-FID), using a Shimadzu GC 2010-Plus (Shimadzu, Kyoto, Japan) equipped with an SP-2560 (100 m × 0.25 mm, 0.20 μm film thickness, Supelco, Bellefonte, PA, USA) capillary column. The injector and detector temperatures were maintained at 220 °C and 250 °C, respectively. The carrier gas was helium at a constant flow of 1 mL/min. The GC oven temperature began at 50 °C for 1 min, then increased to 150 °C at 50 °C/min, held for 20 min, increased to 190 °C at 1 °C/min, and finally increased to 220 °C at 2 °C/min and held for 40 min. The identification of FAMEs was achieved by a comparison of the fatty acid retention times with those of commercial standards (FAME mix, 37 components from

Supelco Inc., Bellefont, PA, USA) and with published chromatograms [18,19]. Additional confirmation of FAMEs and DMAs were achieved by electron impact mass spectrometry using a Shimadzu GC–MS QP2010 Plus (Shimadzu, Kyoto, Japan) equipped with a SP-2560 (100 m × 0.25 mm, 0.20 µm film thickness, Supelco, Bellefonte, PA, USA) capillary column and similar GC conditions.

2.4. Statistical Analysis

FAME and DMA data were analysed as a completely randomised experimental design using the MIXED procedure of SAS 9.4 (SAS Institute Inc., Cary, NC, USA). Diet was used as a fixed factor, and the animal was used as the experimental unit. Feed intake data were analysed as a completely randomised block design, in which an individual lamb was used as the experimental unit and the model included the treatment and initial live weight block as the fixed factors. The least square means and standard error of the mean (SEM) were reported, and the main effects and their interactions were considered significant at $p < 0.05$. The TFA + DMA content is presented in mg/g DM, and the FA individual composition is presented in % of TFA + DMA (g FA/100 g TFA + DMA). The sparse partial least squares discriminant analysis (sPLSDA) was performed using MetaboAnalyst 5.0 software, using the centred log ratio transformed FA data as input.

3. Results

3.1. Fatty Acid Intake

The feed intake averaged 1.19 ± 0.13 kg (mean ± standard error of the mean) of DM/day during the experiment. It did not differ among treatments ($p > 0.05$) [16]. A brief report of individual and total FA intake (mg/day) is presented in Table 2.

Table 2. Daily fatty acid (FA) intake (m/day) during the trial.

FA [1]	Diets [2]				SEM [3]	*p*-Value C vs. *Nannochloropsis* DIETS [4]
	C	O	SD	FD		
14:0	-	320	590	540	0.030	<0.001
16:0	4110	5780	5710	6200	0.466	0.217
*c*9-16:1	60	1300	2490	1970	0.119	<0.001
18:0	660	740	600	790	0.061	0.011
*c*9-18:1	3010	3620	2800	3260	0.277	<0.001
*c*11-18:1	120	230	110	160	0.014	<0.001
18:2n-6	6640	7900	6710	6910	0.607	<0.001
18:3n-3	1440	1700	1190	1530	0.129	<0.001
20:4n-6	-	670	890	750	0.058	<0.001
20:5n-3	-	2600	2710	1630	0.178	<0.001
Total	16,100	25,100	24,000	23,700	1.90	0.033

[1] FA—fatty acid; [2] C—control diet with no EPA sources; O—diet with *Nannochloropsis* sp. oil; SD—diet with spray-dried *Nannochloropsis oceanica* biomass; FD—diet with freeze-dried *Nannochloropsis oceanica* biomass; n.d.—not detected. [3] Standard error of the mean. [4] C vs. *Nannochloropsis* diets, compares C with O, SD and FD together. In the FA notation (x:n-), 'x' represents the number of C atoms, ':' the number of double bonds and 'n-' the location, in its carbon chain, of the double bond which is closest to the methyl end of the molecule. *c* stands for *cis*.

Apart from 16:0, the intake of all the FAs analysed differed between the control and *Nannochloropsis*-diet-supplemented lambs ($p < 0.05$). The FA 14:0, 20:4n-6 and 20:5n-3 were only present in the *Nannochloropsis*-supplemented diets; therefore, their feed intake was zero in the control-fed lambs. The total FA intake differed between control and *Nannochloropsis*-supplemented lambs ($p < 0.05$), being higher in the latter.

3.2. Brain Fatty Acid and Dimethyl Acetal Profile

The total FA and DMA (TFA + DMA) content (mg/g DM) and composition (g/100 TFA + DMA) of grey and white matter are presented in Table 3; those of the hippocampus and prefrontal cortex are presented in Table 4. Regarding the grey and white matter, the amount of TFA + DMA did not differ among treatments, averaging 191 and 227 mg/g DM, respectively. Additionally, in both the hippocampus and prefrontal cortex, TFA + DMA content did not differ among treatments, averaging 199 and 208 mg/g DM, respectively.

Table 3. Total fatty acid (TFA) and dimethyl acetal (DMA) content (mg/g DM) and composition (% TFA + DMA) of the grey matter and white matter of lambs.

FA and DMA [1]	Grey Matter						White Matter					
	Diets [2]				SEM [3]	p-Value	Diets [2]				SEM [3]	p-Value
	C	O	SD	FD			C	O	SD	FD		
TFA + DMA	189	192	190	193	5.2	0.970	235	227	225	222	4.1	0.163
14:0	0.57	0.55	0.58	0.58	0.025	0.746	0.52	0.52	0.55	0.53	0.018	0.778
15:0	0.11	0.12	0.11	0.13	0.008	0.392	0.09	0.09	0.09	0.11	0.007	0.177
16:0	19	19	19	19	0.4	0.942	13	14	14	15	0.4	0.261
*c*7-16:1	0.42	0.42	0.40	0.42	0.021	0.934	0.35	0.37	0.35	0.36	0.016	0.722
*c*9-16:1	0.42	0.43	0.45	0.45	0.024	0.810	0.26	0.29	0.27	0.29	0.013	0.354
17:0	0.29	0.29	0.27	0.31	0.014	0.445	0.27	0.31	0.28	0.31	0.015	0.181
*c*9-17:1	0.11	0.11	0.10	0.11	0.010	0.940	0.11	0.17	0.14	0.15	0.019	0.223
18:0	21	21	20	21	0.5	0.964	15	16	16	16	0.4	0.176
*c*9-18:1	15	15	15	14	0.5	0.847	22	22	22	21	0.5	0.216
*c*11-18:1	3.4	3.3	3.3	3.3	0.07	0.433	3.0	3.0	2.9	3.0	0.05	0.671
18:2n-6	0.48	0.48	0.50	0.52	0.030	0.673	0.29	0.43	0.39	0.42	0.042	0.111
19:1	0.06	0.07	0.07	0.07	0.006	0.836	0.10	0.10	0.10	0.10	0.008	0.892
20:0	0.32	0.31	0.31	0.32	0.017	0.870	0.69 [a]	0.64 [ab]	0.62 [b]	0.62 [b]	0.019	0.045
18:3n-3	0.04	0.03	0.03	0.04	0.008	0.843	0.13 [a]	0.08 [b]	0.13 [a]	0.09 [b]	0.013	0.026
*c*11-20:1	0.86	0.83	0.88	0.80	0.080	0.902	2.2	2.0	2.1	1.9	0.11	0.372
20:1	0.27	0.27	0.27	0.26	0.021	0.941	0.62	0.62	0.59	0.56	0.035	0.542
21:0	0.06	0.06	0.07	0.07	0.005	0.428	0.09	0.10	0.11	0.10	0.012	0.868
20:2	0.10	0.12	0.11	0.12	0.014	0.761	0.26	0.25	0.24	0.23	0.023	0.803
20:2n-6	0.10	0.11	0.11	0.10	0.019	0.959	0.15	0.14	0.14	0.14	0.022	0.638
20:3n-9	0.54	0.54	0.59	0.49	0.068	0.764	0.69	0.69	0.69	0.65	0.054	0.943
22:0	0.71	0.66	0.64	0.66	0.069	0.915	1.90	1.84	1.68	1.69	0.078	0.140
20:3n-6	0.33	0.39	0.36	0.39	0.020	0.094	0.35	0.42	0.38	0.45	0.030	0.111
22:1	0.33	0.32	0.31	0.30	0.040	0.970	0.80	0.10	0.40	0.37	0.170	0.068
20:3n-3	0.13	0.13	0.12	0.11	0.020	0.941	0.57	0.78	0.62	0.60	0.083	0.334
22:1/20:3n-3	0.28	0.20	0.30	0.27	0.064	0.727	0.81	0.88	0.98	0.70	0.067	0.154
20:4n-6	6.3	5.8	5.7	6.2	0.24	0.312	3.8	4.0	3.8	4.2	0.16	0.176
23:0	0.21	0.20	0.21	0.20	0.028	0.992	0.50	0.39	0.39	0.37	0.050	0.262
22:2n-6	0.09	0.09	0.09	0.09	0.018	0.985	0.52	0.48	0.49	0.60	0.119	0.863
20:5n-3	0.26	0.29	0.32	0.32	0.040	0.655	0.71	0.60	0.69	0.69	0.053	0.502
24:0	0.66	0.63	0.61	0.63	0.139	0.997	2.3	2.3	2.2	2.0	0.11	0.115
*c*15-24:1	1.2	1.3	1.4	1.3	0.20	0.951	4.4	3.9	4.2	3.7	0.25	0.251
24:1	0.69	0.20	0.23	0.22	0.200	0.284	0.70	0.65	0.65	0.63	0.038	0.576
22:4n-6	2.5	2.5	2.6	2.9	0.268	0.864	2.8	2.4	2.5	2.7	0.14	0.194
21:5	0.11	0.03	0.03	0.03	0.035	0.334	0.14	0.17	0.12	0.10	0.030	0.448
22:5n-6	0.62	0.41	0.38	0.38	0.066	0.060	0.34	0.25	0.25	0.25	0.037	0.251
26:0	0.15	0.02	0.03	0.04	0.053	0.304	0.08	0.07	0.08	0.08	0.012	0.892
26:1	0.08	0.10	0.13	0.09	0.049	0.878	0.26	0.34	0.42	0.23	0.036	0.069
22:5n-3	3.0	0.76	1.05	1.21	1.111	0.504	0.51	0.32	0.57	0.65	0.150	0.454
22:6n-3	11	14	14	13	1.07	0.248	4.7	5.2	5.2	5.9	0.40	0.212
DMA												
DMA 16:0	2.2	2.2	2.3	2.2	0.10	0.911	3.7	3.6	3.7	3.5	0.09	0.644
DMA 17:0	0.19	0.16	0.16	0.19	0.013	0.320	0.23	0.24	0.23	0.26	0.012	0.319
DMA 18:0	3.9	4.7	4.7	4.7	0.35	0.275	4.5	4.6	4.6	4.7	0.09	0.543
DMA *c*9-18:1	0.92	1.01	1.11	1.01	0.128	0.784	2.8	2.5	2.3	2.5	0.24	0.458
DMA *c*11-18:1	0.95	0.96	1.00	0.95	0.071	0.941	2.0	1.9	2.0	1.9	0.075	0.645

Table 3. *Cont.*

FA and DMA [1]	Grey Matter						White Matter					
	Diets [2]				SEM [3]	p-Value	Diets [2]				SEM [3]	p-Value
	C	O	SD	FD			C	O	SD	FD		
Partial sums												
C18 [4]	39	39	39	39	0.3	0.526	40	41	41	40	0.5	0.207
C18:1 [5]	20	0.27	0.27	0.27	9.97	0.418	0.32	0.34	0.33	0.28	0.027	0.355
DMA	15	9.0	9.2	9.1	2.82	0.341	13	13	13	13	0.2	0.596
SFA [6]	38	42	42	42	1.6	0.270	35	36	35	36	0.6	0.207
MUFA [7]	22	22	22	22	0.8	0.901	35	34	34	32	0.8	0.156
cis-MUFA	22	21	21	21	0.9	0.455	32	32	32	30	0.7	0.223
PUFA [8]	24	26	26	26	0.9	0.519	16	16	16	18	0.6	0.159
n-3 PUFA	14	15	15	15	0.5	0.121	6.7	7.0	7.3	8.0	0.41	0.162
n-6 PUFA	12	9.8	9.8	10	0.55	0.101	8.1	8.0	7.8	8.7	0.33	0.320
EPA + DHA	18	14	14	14	2.2	0.414	5.4	5.8	5.9	6.6	0.41	0.232
AA/EPA ratio	21	23	19	21	3.07	0.815	5.7	6.9	5.7	6.2	0.56	0.375

Means within a row with different letters are significantly different ($p < 0.05$). [1] FA and DMA—fatty acids and dimethyl acetals; [2] C—control diet with no EPA sources; O—diet with *Nannochloropsis* sp. oil; SD—diet with spray-dried *Nannochloropsis oceanica* biomass; FD—diet with freeze-dried *Nannochloropsis oceanica* biomass. [3] Standard error of the mean; the value presented corresponds to a pooled sample standard error of the mean. [4] Sum of C18 FA. [5] Sum of C18:1 FA. [6] Sum of saturated FA. [7] Sum of monounsaturated FA. [8] Sum of polyunsaturated FA. In the FA notation (x:n-), 'x' represents the number of C atoms, ':' the number of double bonds and 'n-' the location, in its carbon chain, of the double bond which is closest to the methyl end of the molecule. *c* stands for *cis*.

Table 4. Total fatty acid (TFA) and dimethyl acetal (DMA) (mg/g DM) content and composition (% TFA + DMA) of the hippocampus and prefrontal cortex of lambs.

FA and DMA [1]	Hippocampus						Prefrontal Cortex					
	Diets [2]				SEM [3]	p-Value	Diets [2]				SEM [3]	p-Value
	C	O	SD	FD			C	O	SD	FD		
TFA + DMA	199	196	204	195	4.4	0.569	206	206	211	210	3.0	0.533
14:0	0.52	0.51	0.53	0.48	0.018	0.243	0.59	0.59	0.61	0.63	0.024	0.643
15:0	0.11	0.10	0.09	0.10	0.005	0.185	0.11	0.11	0.10	0.12	0.008	0.493
16:0	17	17	16	16	0.4	0.458	18	17	17	18	0.3	0.923
*c*7-16:1	0.43	0.43	0.40	0.38	0.017	0.101	0.43	0.42	0.40	0.41	0.018	0.649
*c*9-16:1	0.35	0.39	0.38	0.34	0.017	0.174	0.42	0.42	0.41	0.43	0.019	0.863
17:0	0.29	0.28	0.27	0.29	0.009	0.428	0.27	0.29	0.26	0.31	0.019	0.408
*c*9-17:1	0.11	0.12	0.12	0.12	0.009	0.937	0.13	0.13	0.13	0.15	0.010	0.263
18:0	18	19	18	18	0.4	0.642	19	19	19	19	0.4	0.951
*c*9-18:1	17	17	18	17	0.5	0.756	17	17	17	17	0.5	0.905
*c*11-18:1	3.4	3.3	3.2	3.3	0.05	0.325	3.4	3.3	3.2	3.2	0.05	0.107
18:2n-6	0.41	0.45	0.45	0.44	0.027	0.644	0.40	0.42	0.42	0.45	0.035	0.790
19:1	0.08	0.08	0.08	0.09	0.008	0.578	0.08	0.08	0.07	0.08	0.005	0.125
20:0	0.50	0.48	0.49	0.50	0.029	0.968	0.45	0.48	0.44	0.45	0.027	0.705
18:3n-3	0.09	0.08	0.09	0.06	0.013	0.481	0.07	0.08	0.06	0.06	0.008	0.443
*c*11-20:1	1.4	1.3	1.6	1.6	0.14	0.567	1.2	1.2	1.2	1.2	0.07	0.912
20:1	0.38	0.37	0.40	0.39	0.026	0.822	0.38	0.39	0.38	0.37	0.021	0.878
21:0	0.07	0.07	0.08	0.07	0.007	0.904	0.10	0.09	0.08	0.07	0.013	0.660
20:2	0.19	0.17	0.19	0.17	0.012	0.472	0.19	0.20	0.19	0.18	0.018	0.846
20:2n-6	0.12	0.12	0.12	0.15	0.013	0.297	0.08	0.11	0.10	0.11	0.007	0.075
20:3n-9	0.83	0.84	0.77	0.70	0.051	0.221	0.73	0.74	0.69	0.64	0.066	0.673
22:0	1.0	1.1	1.2	1.2	0.12	0.815	1.1	1.2	1.1	1.2	0.07	0.759
20:3n-6	0.37 [b]	0.44 [a]	0.42 [ab]	0.46 [a]	0.023	0.049	0.29 [b]	0.38 [a]	0.35 [a]	0.37 [a]	0.018	0.008
22:1	0.50	0.47	0.58	0.59	0.054	0.348	0.46	0.53	0.48	0.48	0.032	0.536
20:3n-3	0.24	0.21	0.25	0.24	0.022	0.588	0.26	0.29	0.25	0.27	0.022	0.572
22:1/20:3n-3	0.59	0.49	0.55	0.57	0.145	0.960	0.43	0.30	0.46	0.46	0.037	0.108
20:4n-6	5.4	5.4	4.9	5.1	0.26	0.476	5.3	5.0	4.9	5.1	0.17	0.346
23:0	0.38	0.33	0.36	0.37	0.030	0.641	0.24	0.25	0.24	0.25	0.022	0.980
22:2n-6	0.18	0.13	0.19	0.19	0.035	0.600	0.18	0.20	0.21	0.20	0.038	0.958
20:5n-3	0.56	0.59	0.63	0.62	0.035	0.580	0.55	0.65	0.63	0.65	0.061	0.596
24:0	1.4	1.4	1.5	1.5	0.14	0.971	1.1	1.3	1.2	1.2	0.10	0.805
*c*15-24:1	2.6	2.1	2.7	2.8	0.28	0.266	2.2	2.3	2.3	2.3	0.17	0.983

Table 4. *Cont.*

FA and DMA [1]	Hippocampus						Prefrontal Cortex					
	Diets [2]				SEM [3]	*p*-Value	Diets [2]				SEM [3]	*p*-Value
	C	O	SD	FD			C	O	SD	FD		
24:1	0.39	0.31	0.38	0.40	0.034	0.260	0.43	0.42	0.40	0.44	0.030	0.823
22:4n-6	3.2	2.8	2.8	2.9	0.16	0.235	2.8	2.5	2.6	2.6	0.13	0.329
21:5	0.09	0.08	0.09	0.10	0.014	0.861	0.06	0.07	0.07	0.06	0.008	0.413
22:5n-6	0.50 [a]	0.34 [b]	0.31 [b]	0.31 [b]	0.052	0.047	0.52 [a]	0.38 [ab]	0.34 [b]	0.33 [b]	0.052	0.056
26:0	0.05	0.04	0.05	0.05	0.005	0.286	0.05	0.05	0.05	0.05	0.005	0.675
26:1	0.26	0.18	0.23	0.25	0.067	0.846	0.52	0.35	0.60	0.57	0.078	0.256
22:5n-3	0.83 [c]	1.14 [b]	1.24 [ab]	1.36 [a]	0.066	<0.001	0.54 [c]	0.82 [b]	0.97 [ab]	1.06 [a]	0.069	<0.001
22:6n-3	9.0	9.6	8.9	9.0	0.38	0.573	10	10	10	10	0.53	0.956
DMA												
DMA 16:0	2.7	2.7	2.8	2.7	0.09	0.717	2.7	2.9	2.8	2.8	0.12	0.910
DMA 17:0	0.20	0.20	0.20	0.22	0.013	0.716	0.19	0.19	0.19	0.21	0.013	0.373
DMA 18:0	4.6	4.7	4.7	4.8	0.08	0.612	4.5	4.7	4.7	4.6	0.09	0.519
DMA *c*9-18:1	1.4	1.5	1.8	1.8	0.17	0.268	1.5	1.6	1.6	1.6	0.11	0.986
DMA *c*11-18:1	1.4	1.4	1.5	1.4	0.05	0.301	1.3	1.4	1.4	1.4	0.08	0.919
Partial sums												
C18 [4]	40	40	40	39	0.3	0.303	40	40	40	40	0.3	0.813
C18:1 [5]	0.37	0.32	0.33	0.33	0.018	0.265	0.29	0.28	0.26	0.27	0.019	0.807
DMA	10.3	10	11	11	0.3	0.320	10	11	11	11	0.3	0.830
SFA [6]	39	40	39	39	0.6	0.311	40	40	40	40	0.5	0.934
MUFA [7]	27	26	28	27	0.9	0.617	27	27	27	27	0.7	0.979
cis-MUFA	25	25	26	25	0.8	0.671	25	25	25	25	0.6	0.972
PUFA [8]	22	22	21	22	0.7	0.744	22	22	22	22	0.6	0.952
n-3 PUFA	11	12	11	11	0.3	0.340	12	12	12	12	0.5	0.790
n-6 PUFA	10	9.7	9.2	9.6	0.41	0.362	9.6	9.0	8.9	9.2	0.29	0.309
EPA + DHA	9.6	10	9.5	9.6	0.39	0.600	10	11.0	11	11	0.5	0.960
AA/EPA ratio	9.8	9.2	8.1	8.4	0.72	0.366	10	8.4	8.2	8.3	1.13	0.476

Means within a row with different letters are significantly different ($p < 0.05$). [1] FA and DMA—fatty acids and dimethyl acetals; [2] C—control diet with no EPA sources; O—diet with *Nannochloropsis* sp. oil; SD—diet with spray-dried *Nannochloropsis oceanica* biomass; FD—diet with freeze-dried *Nannochloropsis oceanica* biomass. [3] Standard error of the mean; the value presented corresponds to a pooled sample standard error of the mean. [4] Sum of C18 FA. [5] Sum of C18:1 FA. [6] Sum of saturated FA. [7] Sum of monounsaturated FA. [8] Sum of polyunsaturated FA. In the FA notation (x:n-), 'x' represents the number of C atoms, ':' the number of double bonds and 'n-' the location, in its carbon chain, of the double bond which is closest to the methyl end of the molecule. *c* stands for *cis*.

No major effects of microalgal supplementation ($p > 0.05$) were observed for any of the FAs and DMAs in both the grey and white matter.

Significant differences in the FA composition were observed in the hippocampus and prefrontal cortex (Table 4). In the hippocampus, 20:3n-6 was significantly ($p < 0.05$) lower in the C and SD diets and higher in the O and FD diets. In the prefrontal cortex, this FA was higher in all *Nannochloropsis*-supplemented lambs compared to C-fed lambs. Regarding 22:5n-6: it was significantly higher in the hippocampus in the C-fed lambs when compared to the *Nannochloropsis*-supplemented lambs. In the prefrontal cortex, this FA tended ($p = 0.056$) to follow a similar pattern to what was found in the hippocampus.

DPA had its lowest values ($p < 0.05$) in the hippocampi and prefrontal cortices of C lambs, and its highest values were found in the tissues of lambs fed with the FD supplement. DPA was lower ($p < 0.05$) in these tissues in O-fed lambs than in SD-fed lambs.

Supplement treatments did not affect ($p > 0.05$) the EPA or DHA concentrations in the hippocampus or prefrontal cortex. The sum of EPA + DHA averaged 10% in the hippocampus and 11% in the prefrontal cortex. The total PUFA and n-3 PUFA contents averaged 22% and 12% in both the hippocampus and prefrontal cortex, respectively.

The content of TFA + DMA was not affected ($p > 0.05$) by dietary treatment.

3.3. Retina and Tapetum Lucidum (RTL)

Similar to what was observed in the brain parts, the TFA + DMA content in the RTL tissues did not differ among treatments, averaging 53 mg/g DM (Table 5). However, more treatment effects occurred for RTL tissues than were observed in brain tissues.

In RTL tissues, the *c*16-18:1 value was greater ($p < 0.05$) in SD lambs than in all other treatments. Regarding 20:3n-9, a higher content was found in both the C- and SD-fed lambs when compared to the remaining groups. The content of 20:3n-6 was lower in C-fed lambs when compared to the remaining treatments, and 22:4n-6 was higher in both C- and SD-fed lambs when compared to the remaining treatments. Both EPA and DPA were higher in the microalgae-biomass-fed lambs, although there were no differences between the SD- and FD-fed lambs. When compared to the C treatment, biomass-fed lambs had 4.6 times more EPA and twice more DPA in their RTL tissues.

Table 5. Total fatty acid (TFA) and dimethyl acetal (DMA) (mg/g DM) content and composition (%TFA + DMA) of the RTL tissues of lambs.

FA and DMA [1]	Diet [2]				SEM [3]	*p*-Value
	C	O	SD	FD		
TFA + DMA	54	61	43	53	8.9	0.591
10:0	0.05	0.06	0.08	0.06	0.014	0.376
12:0	0.26	0.24	0.23	0.25	0.036	0.955
14:0	2.2	1.3	1.6	1.9	0.25	0.069
i-15:0	0.08	0.05	0.05	0.02	0.014	0.063
a-15:0	0.18	0.06	0.07	0.05	0.066	0.433
*c*9-14:1	0.06	0.08	0.09	0.07	0.017	0.518
15:0	0.27	0.27	0.28	0.31	0.026	0.766
i-16:0	0.09	0.07	0.10	0.06	0.018	0.341
16:0	21	21	20	21	0.58	0.608
i-17:0	0.15	0.17	0.17	0.15	0.027	0.881
*c*7-16:1	0.29	0.30	0.25	0.29	0.034	0.717
*c*9-16:1	0.77	0.77	0.66	0.83	0.118	0.752
a-17:0	0.18	0.14	0.09	0.13	0.041	0.488
17:0	0.67	0.70	0.60	0.74	0.038	0.084
*c*9-17:1	0.31	0.32	0.27	0.27	0.033	0.543
18:0	18	18	19	18	0.5	0.591
*t*6/*t*7/*t*8-18:1	0.08	0.11	0.11	0.09	0.019	0.664
*t*9-18:1	0.07	0.12	0.11	0.11	0.018	0.198
*t*10-18:1	0.49	0.53	0.28	0.35	0.096	0.219
*t*11-18:1	0.66	0.87	0.88	0.87	0.114	0.455
*t*12-18:1	0.20	0.22	3.56	0.23	1.784	0.437
*c*9-18:1	29	28	21	27	2.4	0.121
*c*11-18:1	1.8	1.8	2.0	1.9	0.10	0.415
*c*12-18:1	0.09	0.11	0.15	0.10	0.023	0.241
*c*13-18:1	0.06	0.06	0.04	0.05	0.014	0.728
*t*16/*c*14-18:1	0.09	0.13	0.10	0.12	0.025	0.671
*c*15-18:1	0.04	0.04	0.05	0.03	0.014	0.799
tc/*ct*-18:2/cyclo-17	0.11	0.18	0.16	0.22	0.043	0.370
*c*9,*t*12/*c*9,*t*15/*t*8,*c*13-18:2	0.09	0.10	0.07	0.11	0.021	0.626
*c*16-18:1	0.03 [b]	0.03 [b]	0.08 [a]	0.04 [b]	0.012	0.021
*t*9,*c*12-18:2	0.04	0.06	0.07	0.05	0.013	0.385
*t*11,*c*15/*t*10,*c*15-18:2	0.06	0.16	0.08	0.14	0.033	0.094
18:2n-6	0.31	3.75	4.25	3.98	0.299	0.317
20:0	0.16	0.17	0.20	0.17	0.022	0.675
18:3n-3	0.63	0.67	0.65	0.66	0.038	0.873
*c*9,*t*11-CLA	0.26	0.37	0.30	0.38	0.057	0.339
20:2n-6	0.15	0.16	0.19	0.17	0.020	0.509
20:3n-9	0.29 [a]	0.20 [bc]	0.24 [ab]	0.14 [c]	0.032	0.014
22:0	0.05	0.07	0.07	0.06	0.012	0.414
20:3n-6	0.38 [b]	0.59 [a]	0.70 [ab]	0.58 [a]	0.056	0.004
20:3n-3	0.18	0.22	0.22	0.20	0.029	0.646
20:4n-6	6.4	5.4	6.5	5.6	0.54	0.391
20:5n-3	0.18 [c]	0.59 [b]	0.83 [a]	0.81 [a]	0.057	<0.001
22:4n-6	0.93 [a]	0.62 [b]	0.73 [ab]	0.58 [b]	0.071	0.008

Table 5. *Cont.*

FA and DMA [1]	Diet [2]				SEM [3]	*p*-Value
	C	O	SD	FD		
22:5n-3	1.2 [c]	2.0 [b]	2.5 [a]	2.4 [ab]	0.17	<0.001
22:6n-3	5.7	6.8	6.9	6.2	0.90	0.752
DMA						
DMA 16:0	1.2	1.2	1.5	1.2	0.14	0.195
DMA 18:0	1.8	1.9	2.1	1.7	0.18	0.552
DMA 18:1	0.41	0.37	0.43	0.34	0.041	0.360
Partial sums						
C18 [4]	37	37	34	36	1.6	0.485
DMA	3.4	3.4	4.0	3.3	0.3	0.351
SFA [5]	42	41	41	42	0.40	0.417
MUFA [6]	34	33	30	32	1.7	0.330
cis-MUFA	32	31	25	30	2.4	0.141
PUFA [7]	20	22	24	22	1.6	0.266
n3-PUFA	7.9	10	11	10	1.0	0.130
n6-PUFA	11	11	12	11	0.9	0.472
EPA + DHA	5.9	7.4	7.7	7.0	0.91	0.491
AA/EPA ratio	37 [a]	9.1 [b]	8.0 [b]	6.9 [b]	2.33	<0.001

Means within a row with different letters are significantly different ($p < 0.05$). [1] FA and DMA—fatty acids and dimethyl acetals; [2] C—control diet with no EPA sources; O—diet with *Nannochloropsis* sp. oil; SD—diet with spray-dried *Nannochloropsis oceanica* biomass; FD—diet with freeze-dried *Nannochloropsis oceanica* biomass. [3] Standard error of the mean; the value presented corresponds to a pooled sample standard error of the mean. [4] Sum of C18 FA. [5] Sum of saturated FA. [6] Sum of monounsaturated FA. [7] Sum of polyunsaturated FA. In the FA notation (x:n-), 'x' represents the number of C atoms, ':' the number of double bonds and 'n-' the location, in its carbon chain, of the double bond which is closest to the methyl end of the molecule. *c* stands for *cis* and *t* stands for *trans*. *i* stands for *iso* and *a* stands for *anteiso*.

The DHA content in RTL tissues did not differ among treatments ($p > 0.05$).

In the partial sums evaluated, the AA/EPA ratio differed between treatments, being higher in the C-fed lambs when compared to the remaining treatments ($p < 0.001$). The sum of the EPA + DHA averaged 7% TFA. N-3 PUFA averaged 10% ± 1.0, corresponding to approximately 45% of the total PUFAs.

Similar to what was verified in the brain, none of the individual DMAs nor the total DMA content differed between treatments in the RTL tissues. Overall, the total DMA content was much lower than that found in the brain, averaging 4%.

The fold change in the EPA, DPA and DHA content between FD-fed lambs and C-fed lambs was compared between the brain, RTL and liver (Table S1) and previously analysed subcutaneous adipose tissue (SC AT) and longissimus lumborum muscle samples [20].

The EPA fold change was higher in the SC AT and similar between the liver and RTL.

3.4. sPLSDA Analysis

Figure 3 illustrates a total of five sparse partial least squares-discriminant analysis (sPLS-DA) plots, corresponding to four plots that belong to all the brain parts evaluated (Panels A–D) and one plot belonging to RTL tissues (Panel E). It is possible to observe that there is no clear individualization of lambs belonging to the same diet in accordance with their brain FA and DMA compositions (%TFA) in all brain parts. However, in the RTL tissues (Figure 3E), it is possible to clearly separate C-fed lambs from *Nannochloropsis*-supplemented lambs based on their retinal FA composition.

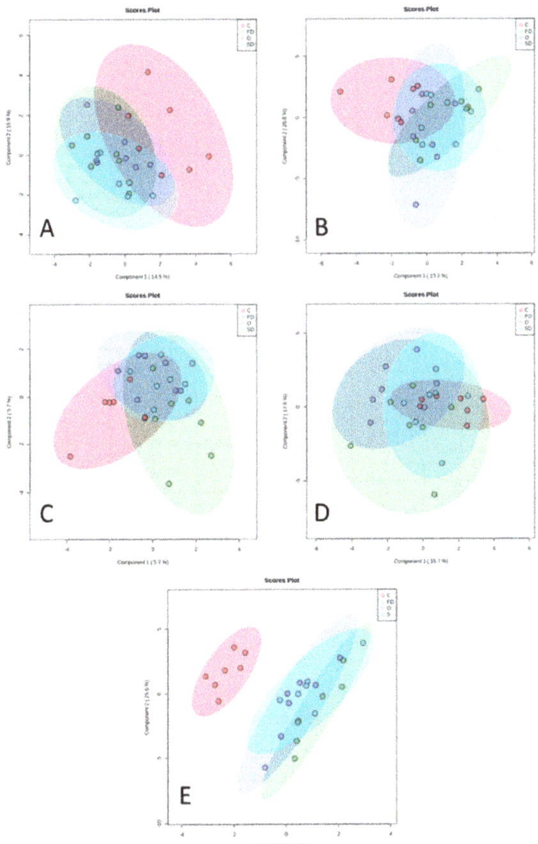

Figure 3. sPLSDA analysis including the FA + DMA profile. The different plots correspond to the different brain parts (**A–D**) and RTL tissues (**E**). (**A**) prefrontal cortex, (**B**) hippocampus, (**C**) grey matter, (**D**) white matter, and (**E**): RTL tissues. Diets are represented in different colours: red—C, control diet with no EPA sources; purple—O, diet with *Nannochloropsis* sp. oil; blue—SD, diet with spray-dried *Nannochloropsis oceanica* biomass; and green—FD, diet with freeze-dried *Nannochloropsis oceanica* biomass.

4. Discussion

4.1. Brain FA Composition

The classic research on ruminant brain lipids has been based on evaluating the lipid classes and the FAs within the lipid classes of whole brain homogenates [10]. In the present study, we present the FA and DMA profile (expressed as % of TFA + DMA) of the brain and RTL tissues of lambs fed *Nannochloropsis* sp. lipids.

The most abundant FAs found in the ovine brain were 18:0 (≈18%), 16:0 (≈17%), *c*9-18:1 (≈17%), DHA (≈9%) and AA (≈5% of TFA + DMA). A similar pattern was also observed in the bovine brain regions [15], with the same five FAs also being the most abundant: 16:0 (≈18%), 18:0 (≈20%), *c*9-18:1 (≈24%), AA (≈6%) and DHA (≈9% of total FA). It was shown that linoleic acid (LA), often the major PUFA in muscle, and α-linolenic acid (ALA) were only present at very low levels. In fact, the longissimus lumborum muscles of the same animals presented LA and ALA proportions of circa 9 and 1% [20], contrasting with the proportions of ≈0.5% and ≈0.1% observed in the brain, respectively. Despite being the major dietary PUFA, LA can hardly be considered functional in the brain because of its low concentration (<0.5% of TFA). This low concentration is probably due to LA being extensively converted to AA, which plays a role in neurodevelopment [21]. Moreover, the majority (≈59%) of LA entering the brain is rapidly β-oxidized [22].

We hypothesised that lambs' brains and/or retinal tissues would be sensitive to the differences in n-3 LC-PUFA absorption due to the changes in rumen biohydrogenation associated with processing the microalgae biomass. Namely, FD *Nannochloropsis oceanica*-supplemented diets appeared to produce a higher n-3 LC-PUFA enhancement in the lambs' brains when compared to O and SD because freeze-drying better protects the integrity of the microalgal cell wall, reducing the access of ruminal microbiota to the n-3 LC-PUFA inside the cell. In ruminants, almost 90% of dietary lipids reach the duodenum as non-esterified saturated FAs [23], and PUFAs are selectively converted into phospholipid forms in the enterocyte [24]. The transport and uptake of EPA and DHA within brain and retina involves their esterification into a lysophospahtidylcholine and a specific transporter (Mfsd2a) [25–27]. Thus we anticipated that the uptake of EPA into the brain and retina would be efficient and responsive to the intestinal absorption of EPA. However, despite the dietary supplementation of EPA, EPA proportions in the brain were low (≈0.6%) and did not differ among treatments in any of the brain parts. This contrasts with the response to EPA deposition observed in the longissimus lumborum muscle of the same animals, in which EPA rose from 0.8% in the C treatment to 1.7% in the *Nannochloropsis*-supplemented treatments [20]. The response in the liver was even more pronounced, with EPA rising from 0.9% up to 4% in the *Nannochloropsis*-supplemented treatments (Table S1).

Thus, in general, the ovine brain was not responsive to dietary EPA supplementation. This contrasts with the results reported by Rule et al. [15], in which a similar intake of EPA (2.3 g/day) resulted in an EPA enhancement across various brain parts. However, the supplementation period in our study lasted 1/10 of the one in Rule et al. [15].

As the uptake of EPA and DHA into the brain is similar [28], the lack of EPA enhancement in the lambs' brains in response to the treatments might be explained by the faster β-oxidation of EPA compared to DHA and/or by the extensive elongation to DPA and subsequent desaturation to DHA. Nevertheless, we also did not observe an increase in DHA. The long half-life of DHA in brain tissues can explain the slow turnover of these fatty acids, therefore explaining the lack of their enhancement in the brain [29]. Moreover, there seems to be evidence that brain DHA and AA levels can be maintained by the liver stores once there is evidence that liver (but not brain) DHA synthesis is upregulated when the dietary content of n-3 PUFA is reduced [13].

Similar responses in brain FAs to microalgae supplementation were observed for the hippocampus and prefrontal cortex, probably reflecting the extensive hippocampal–prefrontal interactions involved in various cognitive and behavioural functions in animals [30,31]. Higher amounts of n-3 DPA and dihomo-γ-linolenic acid (DGLA, 20:3n-6) were found in the hippocampi and prefrontal cortices of microalgae-fed lambs. DGLA is an intermediate of the elongation and desaturation of LA, being converted into AA through the activity of the Δ-5 desaturase enzyme [32]. The increase in DGLA in the brains of lambs supplemented with microalgae was not obvious, as *Nannochloropsis* does not contain relevant amounts of LA and DGLA. DPA, a product of the elongation of EPA, was increased despite the lack of response in EPA and DHA. As it has been proposed that DPA constitutes a storage depot for EPA and DHA [33], its enhancement seems desirable. Although DPA was approximately 10 times lower than DHA in the brain, it was more responsive to the dietary supply of n-3 PUFA. The same pattern was also observed in the brains of lambs suckling from ewes fed with linseed [34] and in the hippocampus of bovines fed fish oil [15].

Most of the beneficial effects of marine oils (mainly fish oils) have been attributed to DHA and EPA [5,35]. However, DPA, which is the intermediate between EPA and DHA in the n-3 LC-PUFA biosynthetic pathway, also presents beneficial biological effects. It reduces platelet aggregation, improves the lipid plasmatic profile, neural health and endothelial cell migration, and assists in the resolution of chronic inflammation [5].

Plasmalogens are a subclass of glycerophospholipids that comprise part of biological membranes, including the plasma membrane and the membranes of intracellular organelles, affecting their biophysical properties. They are quantitively important in membranes of neuronal tissues, including the brain and the retina, and are associated with neurological and psychiatric disorders or are involved in the regulation of retinal vascular development, respectively [36,37]. In the DMA, the backbone at the sn-2 position is mainly bonded to PUFAs such as DHA and AA, suggesting its protective role against lipoxidation [38,39]. The DMA content of ruminant brains is not often reported [40]. In our study, the high abundance of plasmalogens in the brain can be perceived through the high DMA content (≈10% of TFA + DMA). The average content of DMA was higher in the brain when compared to the retina (≈3.5% of TFA + DMA).

4.2. RTL Tissue FA Composition

The retina is a thin, highly organised neural tissue lining the posterior aspect of the eye. It is responsible for initiating vision by transducing light into neural signals [41]. The visual streak area of the retina is a narrow horizontal band. It runs parallel to the ventral edge of the *tapetum* [42]. Therefore, due to anatomical proximity and for practical reasons, we collected both tissues simultaneously. The *tapetum lucidum* is a biologic reflector system that is commonly present in the eyes of vertebrates. It enhances visual sensitivity at low light levels by providing light-sensitive retinal cells with a second opportunity for photon–photoreceptor stimulation. Ovine *tapetum lucidum* belongs to the choroidal fibrous type, and the reflective material is made of collagen [43] that constitutes 65% of the dry weight of the *tapetum* [44]. When comparing the results with the literature, it is important to consider that the specialized retina lipids in the joint RTL samples will be diluted by the fibrous *tapetum* tissue.

Studies with rodents demonstrated that the FA composition of the retina is influenced by diet [45–51] and, for a given species, the retinal FA composition of each phospholipid class is comparable to that of the brain grey matter [6,52]. The retina of sheep and cattle have a similar FA composition [53], and the main FA of bovine retina (% dry weight) are 16:0 (25%), 18:0 (17%), 18:1 (17%) and DHA (23%) [6]. In the present study, 16:0 averaged 20%, 18:0 averaged 18%, *c*9-18:1 averaged 26% and DHA averaged 6.4%. AA averaged 6%, in line with it being the most abundant omega-6 PUFA in the retina [7,8].

Contrary to what was observed in the brain, EPA supplementation increased the EPA content in the RTL tissues. In the C lambs, EPA averaged 0.18% TFA + DMA, which was in line with human EPA retinal content [54]. The EPA content significantly increased in O-fed lambs (0.59% TFA + DMA) and particularly in SD- and FD-fed lambs (0.82% TFA + DMA; + 4.6 times the EPA content in the C-fed lambs). Contrary to the results achieved in the brain, our results showed that the RTL tissues of lambs are very responsive to EPA supplementation. The same magnitude of response was only comparable to what we found in the liver (Figure 4).

The 4.6-fold increase was achieved despite the short duration of EPA supplementation. Consistent with a better responsiveness of RTL tissues to the experimental diets, control lambs were clearly separated from the lambs consuming *Nannochloropsis*-supplemented diets in the sPLSDA analysis. This shows, once again, that the RTL tissues seem to have been much more sensitive to dietary intervention. The high responsiveness of the retina is evident in rodent studies, in which EPA contents of 6 to 35 times greater have been reported following EPA supplementation [26,55].

As in the brain, no alterations in DHA content were observed between different treatments. The content of DHA in RTL tissues averaged 6.4% of the total TFA + DMA. This is considerably lower that what was reported in previous studies for ruminants in which the DHA content averaged approximately 20–30% [6,53,56].

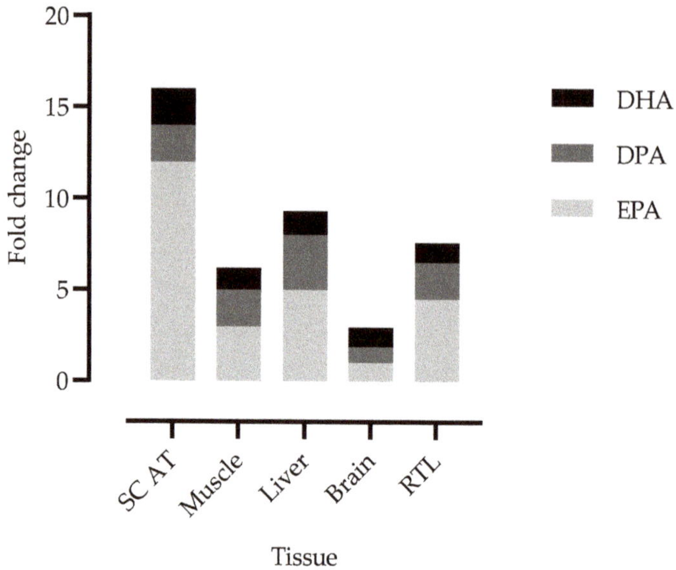

Figure 4. EPA, DPA and DHA fold change in lambs' tissues. The fold change was calculated between the mean value determined in the tissues of freeze-dried *Nannochloropsis oceanica*-fed lambs versus the mean value for control-fed lambs. SC AT: subcutaneous adipose tissue; RTL: retina and *tapetum lucidum*. The reference value for the brain corresponds to the mean of all brain parts.

5. Conclusions

After a short-term trail of EPA supplementation in lambs, achieved through feeding using three different diets containing *Nannochloropsis* sp. microalga, it was possible to conclude that the brain content of EPA was not responsive to dietary supplementation. However, the EPA content in the retina was highly responsive in lambs supplemented with *Nannochloropsis*, especially lambs consuming SD and FD diets. Although we could not confirm an advantage in freeze-drying over spray-drying *Nannochloropsis oceanica* with respect to the efficiency of EPA enhancement in the lambs' retinal tissues, we can confirm their advantage over the free oil. Overall, our results suggest that RTL is a good target to evaluate the differences in n-3 LC-PUFA absorption due to the changes in rumen biohydrogenation associated with dietary interventions.

Supplementary Materials: The following supporting information can be downloaded at: https://www.mdpi.com/article/10.3390/ani13050828/s1, Table S1: Summary table for total fatty acid (TFA) and dimethyl acetal (DMA) (mg/g DM) content and composition (%TFA + DMA) of the liver tissues of lambs.

Author Contributions: S.P.A. and R.J.B.B., were responsible for conceptualization and funding acquisition; A.C.M.V., R.J.B.B. and S.P.A., implemented the experiment; A.C.M.V. conducted the animal experiment; A.C.M.V. and S.P.A. conducted laboratory analysis; J.J.C. monitored and assisted histopathological analyses; A.C.M.V., R.J.B.B. and S.P.A. performed data analysis; A.C.M.V., R.J.B.B. and S.P.A. interpreted the results; A.C.M.V. drafted the manuscript; and J.J.C., S.P.A. and R.J.B.B. edited and revised the original draft. All authors have read and agreed to the published version of the manuscript.

Funding: This research was funded by Fundação para a Ciência e Tecnologia (FCT) through the PhD grant SFRH/BD/136609/2018 awarded to ACM Vítor, the research contract to S.P.A. (DL 57/2016/CP1438/CT0008) and the research projects PTDC/CAL-ZOO/29654/2017 and UIDB/00276/2020 (CIISA) and LA/P/0059/2020 (AL4AnimalS).

Institutional Review Board Statement: The experimentation involving live animals was conducted under strict compliance with international guidelines (Directive 2010/63/EU) regulating the use of production animals in animal experimentation. Experimentation was conducted at the INIAV-Santarém facilities. The INIAV-Santarém facilities are certified by the competent veterinary authority (DGAV) to conduct animal experimentation (Ref: 04211000/000/2013). The experimental animal procedures were approved by the Ethical and Animal Well-Being Commission (CEBEA) of the Faculty of Veterinary Medicine, University of Lisbon, Portugal (Protocol FMV/CEBEA 007/2016). Animal management, handling, transport, and sacrifice were conducted replicating approved standard commercial practices regarding animal welfare except that animals were individually housed. The study was carried out in compliance with the ARRIVE guidelines.

Informed Consent Statement: Not applicable.

Data Availability Statement: The data presented in this study are available on request from the corresponding author.

Acknowledgments: The authors acknowledge the support from Fundação para a Ciência e Tecnologia (FCT) through the PhD grant SFRH/BD/136609/2018 awarded to ACM Vítor, the research contract to S.P.A. (DL 57/2016/CP1438/CT0008) and the research projects PTDC/CAL-ZOO/29654/2017, UIDB/00276/2020 and UIDB/05183/2020. The authors acknowledge the collaboration of Instituto Superior de Agronomia, University of Lisbon, for helping with the microalgae processing and helping with the management of the animals. The authors also acknowledge the INIAV staff for the support in animal management and slaughter.

Conflicts of Interest: The authors declare no conflict of interest.

References

1. Svennerholm, L. Distribution and fatty acid composition of phosphoglycerides in normal human brain. *J. Lipid Res.* **1968**, *9*, 570–579. [CrossRef] [PubMed]
2. Fenton, W.S.; Hibbeln, J.; Knable, M. Essential fatty acids, lipid membrane abnormalities, and the diagnosis and treatment of schizophrenia. *Biol. Psychiatry* **2000**, *47*, 8–21. [CrossRef] [PubMed]
3. Lange, K. Omega-3 fatty acids and mental health. *Glob. Health J.* **2020**, *4*, 18–30. [CrossRef]
4. Dyall, S.C. Long-chain omega-3 fatty acids and the brain: A review of the independent and shared effects of EPA, DPA and DHA. *Front. Aging Neurosci.* **2015**, *7*, 52. [CrossRef]
5. Drouin, G.; Rioux, V.; Legrand, P. The n-3 docosapentaenoic acid (DPA): A new player in the n-3 long chain polyunsaturated fatty acid family. *Biochimie* **2019**, *159*, 36–48. [CrossRef]
6. Fliesler, S.J.; Anderson, R.E. Chemistry and metabolism of lipids in the vertebrate retina. *Prog. Lipid Res.* **1983**, *22*, 79–131. [CrossRef]
7. Acar, N.; Berdeaux, O.; Gregoire, S.; Cabaret, S.; Martine, L.; Gain, P.; Thuret, G.; Creuzot-Garcher, C.P.; Bron, A.M.; Bretillon, L. Lipid composition of the human eye: Are red blood cells a good mirror of retinal and optic nerve fatty acids? *PLoS ONE* **2012**, *7*, e35102. [CrossRef]
8. Albouery, M.; Buteau, B.; Grégoire, S.; Martine, L.; Gambert, S.; Bron, A.M.; Acar, N.; Chassaing, B.; Bringer, M.A. Impact of a high-fat diet on the fatty acid composition of the retina. *Exp. Eye Res.* **2020**, *196*, 108059. [CrossRef]
9. Querques, G.; Forte, R.; Souied, E.H. Retina and omega-3. *J. Nutr. Metab.* **2011**, *2011*, 748361. [CrossRef]
10. Christie, W. *Lipid Metabolism in Ruminant Animals*; Pergamon Press: Oxford, UK, 1981; Volume 2, pp. 95–191.
11. Anderson, R.; Sperling, L. Lipids of ocular tissues. VII Positional distribution of the fatty acids in the phospholipids of bovine retina rod outer segments. *Arch. Biochem. Biophys.* **1971**, *144*, 1163–1168.
12. Ponnampalam, E.; Sinclair, A.; Holman, B. The sources, synthesis and biological actions of omega-3 and omega-6 fatty acids in red meat: An overview. *Foods* **2021**, *10*, 1358. [CrossRef]
13. Rapoport, S.I.; Igarashi, M.; Gao, F. Quantitative contributions of diet and liver synthesis to docosahexaenoic acid homeostasis. *Prostaglandins Leukot. Essent. Fat. Acids* **2010**, *82*, 273–276. [CrossRef]
14. Igarashi, M.; DeMar, J.C.; Ma, K.; Chang, L.; Bell, J.M.; Rapoport, S.I. Upregulated liver conversion of α-linolenic acid to docosahexaenoic acid in rats on a 15 week n-3 PUFA-deficient diet. *J. Lipid Res.* **2007**, *48*, 152–164. [CrossRef]
15. Rule, D.C.; Melson, E.A.; Alexander, B.M.; Brown, T.E. Dietary Fatty Acid Composition Impacts the Fatty Acid Profiles of Different Regions of the Bovine Brain. *Animals* **2022**, *12*, 2696. [CrossRef]
16. Vítor, A.C.M.; Francisco, A.E.; Silva, J.; Pinho, M.; Huws, S.A.; Santos-Silva, J.; Bessa, R.J.B.; Alves, S.P. Freeze-dried *Nannochloropsis oceanica* biomass protects eicosapentaenoic acid (EPA) from metabolization in the rumen of lambs. *Sci. Rep.* **2021**, *11*, 1–16. [CrossRef]
17. Christie, W.; Han, X. Chapter 7—Preparation of Derivatives of Fatty Acids. In *Lipid Analysis*, 4th ed.; Woodhead Publishing Limited: Cambridge, UK, 2012; pp. 145–158.

18. Alves, S.P.; Bessa, R.J.B. Comparison of two gas-liquid chromatograph columns for the analysis of fatty acids in ruminant meat *J. Chromatogr. A* **2009**, *1216*, 5130–5139. [CrossRef]
19. Alves, S.; Santos-Silva, J.; Cabrita, A.R.J.; Fonseca, A.J.M.; Bessa, R.J.B. Detailed dimethylacetal and fatty acid composition of rumen content from lambs fed lucerne or concentrate supplemented with soybean oil. *PLoS ONE* **2013**, *8*, e58386. [CrossRef]
20. Vítor, A.; Godinho, M.; Francisco, A.E.; Silva, J.; Almeida, J.; Fialho, L. *Nannochloropsis oceanica* microalga feeding increases long-chain omega-3 polyunsaturated fatty acids in lamb meat. *Meat Sci.* **2023**, *197*, 109053. [CrossRef]
21. Taha, A.Y. Linoleic acid—good or bad for the brain ? *NPJ Sci. Food* **2020**, *4*, 1. [CrossRef]
22. DeMar, J.C., Jr.; Lee, H.J.; Ma, K.; Chang, L.; Bell, J.M.; Rapoport, S.I.; Bazinet, R.P. Brain elongation of linoleic acid is a negligible source of the arachidonate in brain phospholipids of adult rats. *Biochim. Biophys. Acta (BBA)-Mol. Cell Biol. Lipids* **2006**, *1761*, 1050–1059. [CrossRef]
23. Doreau, M.; Ferlay, A. Digestion and utilisation of fatty acids by ruminants. *Anim. Feed. Sci. Technol.* **1994**, *45*, 379–396. [CrossRef]
24. Moore, J.; Christie, W. Digestion, absorption and transport of fats in ruminant animals. In *Fats in Animal Nutrition*, 1st ed.; Wiseman, J., Ed.; Butterworth-Heinemann: Oxford, UK, 1984; pp. 123–149.
25. Nguyen, L.N.; Ma, D.; Shui, G.; Wong, P.; Cazenave-Gassiot, A.; Zhang, X.; Wenk, M.R.; Goh, E.L.K.; Silver, D.L. Mfsd2a is a transporter for the essential omega-3 fatty acid docosahexaenoic acid. *Nature* **2014**, *509*, 503–506. [CrossRef] [PubMed]
26. Yalagala, P.C.R.; Sugasini, D.; Dasarathi, S.; Pahan, K.; Subbaiah, P.V. Dietary lysophosphatidylcholine-EPA enriches both EPA and DHA in the brain: Potential treatment for depression. *J. Lipid Res.* **2019**, *60*, 566–578. [CrossRef] [PubMed]
27. Sugasini, D.; Yalagala, P.C.R.; Subbaiah, P.V. Efficient enrichment of retinal DHA with dietary lysophosphatidylcholine-DHA: Potential application for retinopathies. *Nutrients* **2020**, *12*, 3114. [CrossRef]
28. Chen, C.T.; Bazinet, R.P. β-oxidation and rapid metabolism, but not uptake regulate brain eicosapentaenoic acid levels *Prostaglandins Leukot. Essent. Fat. Acids* **2015**, *92*, 33–40. [CrossRef]
29. Rapoport, S.I. In vivo fatty acid incorporation into brain phospholipids in relation to plasma availability, signal transduction and membrane remodeling. *J. Mol. Neurosci.* **2001**, *16*, 243–261. [CrossRef]
30. Sigurdsson, T.; Duvarci, S. Hippocampal-prefrontal interactions in cognition, behavior and psychiatric disease. *Front. Syst. Neurosci.* **2016**, *9*, 190. [CrossRef]
31. Eichenbaum, H. Prefrontal-hippocampal interactions in episodic memory. *Nat. Rev. Neurosci.* **2017**, *18*, 547–558. [CrossRef]
32. Wang, X.; Lin, H.; Gu, Y. Multiple roles of dihomo-γ-linolenic acid against proliferation diseases. *Lipids Health Dis.* **2012**, *11*, 25. [CrossRef]
33. Miller, E.; Kaur, G.; Larsen, A.; Loh, S.P.; Linderborg, K.; Weisinger, H.S.; Turchini, G.M.; Cameron-Smith, D.; Sinclair, A.J. A short-term n-3 DPA supplementation study in humans. *Eur. J. Nutr.* **2013**, *52*, 895–904. [CrossRef]
34. Nudda, A.; Bee, G.; Correddu, F.; Lunesu, M.F.; Cesarani, A.; Rassu, S.P.G.; Pulina, G.; Battacone, G. Linseed supplementation during uterine and early post-natal life markedly affects fatty acid profiles of brain, liver and muscle of lambs. *Ital. J. Anim. Sci.* **2022**, *21*, 361–377. [CrossRef]
35. FAO/WHO. Interim Summary of Conclusions and Dietary Recommendations on Total Fat & Fatty Acids. *Jt FAO/WHO Expert Consult Fats Fat. Acids Hum. Nutr.* **2008**.
36. Saab, S.; Mazzocco, J.; Creuzot-garcher, C.P. Plasmalogens in the retina : From occurrence in retinal cell membranes to potential involvement in pathophysiology of retinal diseases. *Biochimie* **2014**, *107*, 58–65. [CrossRef]
37. Udagawa, J.; Hino, K. Plasmalogen in the brain: Effects on cognitive functions and behaviors attributable to its properties. *Brain Res. Bull.* **2022**, *1*, 197–202. [CrossRef]
38. Braverman, N.E.; Moser, A.B. Functions of plasmalogen lipids in health and disease. *Biochim. Biophys. Acta (BBA) Mol. Basis Dis.* **2012**, *1822*, 1442–1452. [CrossRef]
39. Dorninger, F.; Forss-Petter, S.; Wimmer, I.; Berger, J. Plasmalogens, platelet-activating factor and beyond—Ether lipids in signaling and neurodegeneration. *Neurobiol. Dis.* **2020**, *145*, 105061. [CrossRef]
40. Alfaia, C.M.; Alves, S.P.; Pestana, J.M.; Madeira, M.S.; Moreira, O.; Santos-Silva, J.; Bessa, R.J.; Toldrá, F.; Prates, J.A. Distinct fatty acid composition of some edible by-products from bovines fed high or low silage diets. *Food Sci. Technol. Int.* **2017**, *23*, 209–221. [CrossRef]
41. Smith, J.D.; Greenlee, J.J.; Hamir, A.N.; Richt, J.A.; Greenlee, M.H.W. Retinal function and morphology are altered in cattle infected with the prion disease transmissible mink encephalopathy. *Veter Pathol.* **2009**, *46*, 810–818. [CrossRef]
42. Lossi, L. Anatomical features for an adequate choice of the experimental animal model in biomedicine: III. Ferret, goat, sheep, and horse. *Ann. Anat. Anat. Anz.* **2022**, *244*, 151978. [CrossRef]
43. Ollivier, F.J.; Samuelson, D.A.; Brooks, D.E.; Lewis, P.A.; Kallberg, M.E.; Komáromy, A.M. Comparative morphology of the tapetum lucidum (among selected species). *Vet. Ophthalmol.* **2004**, *7*, 11–22. [CrossRef]
44. Bellairs, R.; Harkness, M.L.R.; Harkness, R.D. The structure of the tapetum of the eye of the sheep. *Cell Tissue Res.* **1975**, *157*, 73–91. [CrossRef] [PubMed]
45. Chong, E.W.T.; Kreis, A.J.; Wong, T.Y.; Simpson, J.A.; Guymer, R.H. Dietary ω-3 fatty acid and fish intake in the primary prevention of age-related macular degeneration: A systematic review and meta-analysis. *Arch Ophthalmol.* **2008**, *126*, 826–833. [CrossRef] [PubMed]

46. Merle, B.; Delyfer, M.N.; Korobelnik, J.F.; Rougier, M.B.; Colin, J.; Malet, F.; Féart, C.; Le Goff, M.; Dartigues, J.F.; Barberger-Gateau, P.; et al. Dietary omega-3 fatty acids and the risk for age-related maculopathy: The alienor study. *Investig. Ophthalmol. Vis. Sci.* **2011**, *52*, 6004–6011. [CrossRef] [PubMed]
47. Schnebelen, C.; Viau, S.; Grégoire, S.; Joffre, C.; Creuzot-Garcher, C.P.; Bron, A.M.; Bretillon, L.; Acar, N. Nutrition for the eye: Different susceptibility of the retina and the lacrimal gland to dietary omega-6 and omega-3 polyunsaturated fatty acid incorporation. *Ophthalmic. Res.* **2009**, *41*, 216–224. [CrossRef]
48. Schnebelen, C.; Grégoire, S.; Pasquis, B.; Joffre, C.; Creuzot-Garcher, C.P.; Bron, A.M.; Bretillon, L.; Acar, N. Dietary n-3 and n-6 PUFA enhance DHA incorporation in retinal phospholipids without affecting PGE1 and PGE2 levels. *Lipids* **2009**, *44*, 465–470. [CrossRef]
49. Seddon, J.M.; Rosner, B.; Sperduto, R.D.; Yannuzzi, L.; Haller, J.A.; Blair, N.P.; Willett, W. Dietary fat and risk for advanced age-related macular degeneration. *Arch Ophthalmol.* **2001**, *119*, 1191–1199. [CrossRef]
50. Simon, E.; Bardet, B.; Grégoire, S.; Acar, N.; Bron, A.M.; Creuzot-Garcher, C.P.; Bretillon, L. Decreasing dietary linoleic acid promotes long chain omega-3 fatty acid incorporation into rat retina and modifies gene expression. *Exp. Eye Res.* **2011**, *93*, 628–635. [CrossRef]
51. Tan, J.S.L.; Wang, J.J.; Flood, V.; Mitchell, P. Dietary fatty acids and the 10-year incidence of age-related macular degeneration: The blue mountains eye study. *Arch Ophthalmol.* **2009**, *127*, 656–665. [CrossRef]
52. Bartley, W.; Van Heyningen, R.; Notton, B.M.; Renshaw, A. Fatty acid composition of lipids present in different parts of the ox eye. *Biochem. J.* **1962**, *85*, 332–335. [CrossRef]
53. Anderson, R.E. Lipids of ocular tissues. IV. A comparison of the phospholipids from the retina of six mammalian species. *Exp. Eye Res.* **1970**, *9*, 281–284. [CrossRef]
54. Bretillon, L.; Thuret, G.; Grégoire, S.; Acar, N.; Joffre, C.; Bron, A.M.; Gain, P.; Creuzot-Garcher, C.P. Lipid and fatty acid profile of the retina, retinal pigment epithelium/choroid, and the lacrimal gland, and associations with adipose tissue fatty acids in human subjects. *Exp. Eye Res.* **2008**, *87*, 521–528. [CrossRef]
55. Shindou, H.; Koso, H.; Sasaki, J.; Nakanishi, H.; Sagara, H.; Nakagawa, K.M.; Takahashi, Y.; Hishikawa, D.; Iizuka-Hishikawa, Y.; Tokumasu, F.; et al. Docosahexaenoic acid preserves visual function by maintaining correct disc morphology in retinal photoreceptor cells. *J. Biol. Chem.* **2017**, *292*, 12054–12064. [CrossRef]
56. Lecomte, M.; Paget, C.; Ruggiero, D.; Wiernsperger, N.; Lagarde, M. Docosahexaenoic acid is a major n-3 polyunsaturated fatty acid in bovine retinal microvessels. *J. Neurochem.* **1996**, *66*, 2160–2167. [CrossRef]

MDPI
St. Alban-Anlage 66
4052 Basel
Switzerland
Tel. +41 61 683 77 34
Fax +41 61 302 89 18
www.mdpi.com

Animals Editorial Office
E-mail: animals@mdpi.com
www.mdpi.com/journal/animals

www.ingramcontent.com/pod-product-compliance
Lightning Source LLC
Chambersburg PA
CBHW062036220926
51850CB00010B/2569

9783036571546